Joint AIChE/SPE Workshop: Next Generation Deepwater Facilities 2016

Houston, Texas, USA
12 – 14 December 2016

ISBN: 978-1-5108-6373-6

Printed from e-media with permission by:

Curran Associates, Inc.
57 Morehouse Lane
Red Hook, NY 12571

Printed by Curran Associates, Inc. (2018)

AIChE
120 Wall Street, FL 23
New York, NY 10005-4020

Phone: (800) 242-4363
Fax: (203) 775-5177

www.aiche.org

Additional copies of this publication are available from:

Curran Associates, Inc.
57 Morehouse Lane
Red Hook, NY 12571 USA
Phone: 845-758-0400
Fax: 845-758-2633
Email: curran@proceedings.com
Web: www.proceedings.com

TABLE OF CONTENTS

Next Generation Deepwater Facilities

A View Forward Based on the Past

by
Ken Arnold
K Arnold Consulting, Inc.

AIChE
The Global Home of Chemical Engineers

Joint AIChE/SPE Workshop: Next Generation Deepwater Facilities
December 12-14, 2016 | Hyatt Regency Houston/Galleria | Houston, TX

Society of Petroleum Engineers

OUTLINE

Understanding the Problem

Lessons from the Past

- Subsea
- Floater
- Topsides
- Project Management

Thoughts for a Way Forward

AIChE
The Global Home of Chemical Engineers

Joint AIChE/SPE Workshop: Next Generation Deepwater Facilities
December 12-14, 2016 | Hyatt Regency Houston/Galleria | Houston, TX

Society of Petroleum Engineers

MAKING A PROJECT VIABLE

It is not just lowering capital cost

- Operating cost
- Safety and Reliability
- Cash flow

Evaluating incremental increase in investment

- $200MM yields PVP $300MM
- $100MM yields PVP $275MM

Putting the facilities capital cost in perspective

- Facilities capital cost vs. PV cash flow from production

Are we spending too much time and money on facilities FEEDS?

WHAT DO WE NEED?

New technology

Standardization

Proper use of old technology

Appropriate commercial plan of activity

Elimination of studies to prove the obvious

Make decisions

- Accept we don't know everything we would like to know
- Simplicity in design
- Flexibility vs. process optimization
- Reduce focus on long term optimization
- Start small and rejoice in the upside if it happens

AIChE The Global Home of Chemical Engineers
Joint AIChE/SPE Workshop: Next Generation Deepwater Facilities
December 12-14, 2016 | Hyatt Regency Houston/Galleria | Houston, TX
SPE International Society of Petroleum Engineers

OUTLINE

Understanding the Problem

Lessons from the Past

- Subsea
- Floater
- Topsides
- Project Management

Thoughts for a Way Forward

AIChE The Global Home of Chemical Engineers

Joint AIChE/SPE Workshop: Next Generation Deepwater Facilities
December 12-14, 2016 | Hyatt Regency Houston/Galleria | Houston, TX

SPE International Society of Petroleum Engineers

SHELL LOCKHEED – 1960s

Task:

- Complete subsea wells in up to 1500 feet of water
- Manifold them and provide for testing

Problem:

- Capabilities of ROVs
- Instrumentation and controls reliability
- No multiphase flow meters

Solution:

- One atmosphere chambers with diving bell access
- Wellhead- 1972
- Manifold and test separator- 1975

Result:

- Successful program but wrong solution

AIChE The Global Home of Chemical Engineers

Joint AIChE/SPE Workshop: Next Generation Deepwater Facilities
December 12-14, 2016 | Hyatt Regency Houston/Galleria | Houston, TX

SPE International Society of Petroleum Engineers

SUBSEA PROCESSING TO REDUCE TOPSIDE WEIGHT

Task:

- Separate oil, water and gas subsea
- Dispose of water subsea
- Reduce flow assurance and transport costs

Problem:

- Treat the oil and dispose of water
- Treat or re-inject the gas
- Provide support and utilities to subsea

Results:

- Working the problem for 25 years
- Need to demonstrate long term operating, maintenance and reliability

SUBSEA PROCESSING TO REDUCE TOPSIDE WEIGHT (CONTINUED)

Question: Do We Understand the Savings

Long tiebacks

- Hydrates, paraffins, asphaltenes
- Backpressure on wells

Do we Really Eliminate Topside Cost & Weights

- Analysis of 75,000 BOPD GOM topsides weight savings:
  - Gas/Liquid Separation: 2.6%
  - Bulk Oil/Water Separation: 2.5%
  - Oil Treating: 8.0%
  - Water Treating: 2.2%

OUTLINE

Understanding the Problem

Lessons from the Past

- Subsea
- Floater
- Topsides
- Project Management

Thoughts for a Way Forward

AIChE The Global Home of Chemical Engineers

Joint AIChE/SPE Workshop: Next Generation Deepwater Facilities
December 12-14, 2016 | Hyatt Regency Houston/Galleria | Houston, TX

Society of Petroleum Engineers

ONE GOM OPERATOR

Business Model

- Develop marginal fields
- Minimal capital cost- Mini-TLP

Problem

- Fields others didn't want
- Limited upside potential
- Reservoir risk

Results

- Reservoir problems
- Slight cost overruns
- Bankruptcy

AIChE The Global Home of Chemical Engineers

Joint AIChE/SPE Workshop: Next Generation Deepwater Facilities
December 12-14, 2016 | Hyatt Regency Houston/Galleria | Houston, TX

Society of Petroleum Engineers

TYPHOON

AIChE The Global Home of Chemical Engineers

Joint AIChE/SPE Workshop: Next Generation Deepwater Facilities
December 12-14, 2016 | Hyatt Regency Houston/Galleria | Houston, TX

SPE International Society of Petroleum Engineers

CONSIDERATIONS FOR NEXT GENERATION

Upside potential is important

Elegant solutions to reduce cost may introduce risk

AIChE The Global Home of Chemical Engineers

Joint AIChE/SPE Workshop: Next Generation Deepwater Facilities
December 12-14, 2016 | Hyatt Regency Houston/Galleria | Houston, TX

SPE International Society of Petroleum Engineers

OUTLINE

Understanding the Problem

Lessons from the Past

- Subsea
- Floater
- Topsides
- Project Management

Thoughts for a Way Forward

FIT FOR PURPOSE

Four aspects cause cost differences:

- Commercial plan of execution
- Documentation required for project management
- Quality standards
- Design choices and complexity

Commercial plan of execution is often dictated by country

Documentation often dictated by regulatory regime and oil company policies

Quality standards alone have some impact but normally of the order of 10 to 25% of equipment cost (4 to 10% of topside cost)

Design choices and complexity have greatest impact

STANDARD FACILITIES

Shell, late 1960s

- Built 3 complete sets of standard modules for GOM 8-pile decks
- Installed one; heavily modified the other two
- Not practical

Natco, mid 1970s

- Built several standard 50 MMSCFD topsides for 4-pile decks
- Not a viable sales tool

Problem: There are no ASTM standard reservoirs

SHELL TYPICAL FACILITIES MID 1970s TO MID 1980s

No two reservoirs and fields are exactly alike

- It is not possible to design a standard facility

Facilities are made up almost entirely of similar building blocks

- It is possible to develop a set of fit for purpose quality specifications for 90+% of needs
- It is possible to develop a standard process for making overall design decisions
- It is possible to develop a set of unsized "typical facility", fit for purpose building block P&IDs

AIChE The Global Home of Chemical Engineers
Joint AIChE/SPE Workshop: Next Generation Deepwater Facilities
December 12-14, 2016 | Hyatt Regency Houston/Galleria | Houston, TX
Society of Petroleum Engineers

CONSIDERATIONS FOR NEXT GENERATION

Advances in separation and treating equipment are helpful

An agreed industry design for a "standard" topside will probably not work

Individual companies could adopt "typical facilities" concepts

OUTLINE

Understanding the Problem

Lessons from the Past

- Subsea
- Floater
- Topsides
- Project Management

Thoughts for a Way Forward

NS VS GOM PM PHILOSOPHIES

Paragon 1989 OTC:

- 60 to 120% increase due to design philosophy

1990 BP report by Davy McDermott:

- Controllable topsides premium 70%

UK Department of Trade and Industry 1999:

- Incorporating GOM practices can save roughly $100MM (~80% premium)
- GOM differs from NS in project execution and philosophy:
  - Oil companies act as prime contractors
  - Regulatory system requires less documentation
  - Less instrument complexity
  - Use of standard vendor packages with ancillary items
  - Fewer safety studies
  - HA based on experience captured in codes and judgment
  - Specification and quality documentation
  - Fewer man-hours for inspection, project auditing and control services
  - More engineering consistency from job to job

GOM TOPSIDES INDEPENDENT VS. MAJOR OIL COMPANY 1999

48% Premuim

	Independent	Major
BOPD	60,000	60,000
MMSCFD	200	120
BWPD	75,000	30,000
Water Injection, BWPD	Future (90,000)	72,000
Construction type	Skids	Integrated
Topside TIC	$64 MM	$145 MM
TIC adjusted for compression and waterflood	$64 MM	$95 MM

AIChE The Global Home of Chemical Engineers
Joint AIChE/SPE Workshop: Next Generation Deepwater Facilities
December 12-14, 2016 | Hyatt Regency Houston/Galleria | Houston, TX
SPE International Society of Petroleum Engineers

COMPANY "X"
70 MMSCFD GLYCOL DEHY BID PACKAGE

Bid package size:

- "X" GOM: 12-inches
- Paragon WA: 150 pages
- Independent GOM: 16 pages

Results:

- One bidder declined
- One bidder read specs and was 50% high
- Successful bidder applied company factor
- "Contract inspectors don't know what is in there"
- 43 to 63% higher cost than Paragon/Independent

CONSIDERATIONS FOR NEXT GENERATION

Large savings are possible based on project design philosophies and commercial plan of execution

Questions:

- Are we over-analyzing the design?
- Is all the documentation really adding value?
- What is the right level of instrumentation?
- Are all the safety studies really needed?
- Can "typical facilities" help with engineering consistency and removing complexity?

AIChE The Global Home of Chemical Engineers
Joint AIChE/SPE Workshop: Next Generation Deepwater Facilities
December 12-14, 2016 | Hyatt Regency Houston/Galleria | Houston, TX
SPE International Society of Petroleum Engineers

OUTLINE

Understanding the Problem

Lessons from the Past

- Subsea
- Floater
- Topsides
- Project Management

Thoughts for a Way Forward

Joint AIChE/SPE Workshop: Next Generation Deepwater Facilities
December 12-14, 2016 | Hyatt Regency Houston/Galleria | Houston, TX

WHAT DO WE NEED?

New technology

Standardization

Proper use of old technology

Appropriate commercial plan of activity

Elimination of studies to prove the obvious

Make decisions

- Accept we don't know everything we would like to know
- Simplicity in design
- Flexibility vs. process optimization
- Reduce focus on long term optimization
- Start small and rejoice in the upside if it happens

Joint AIChE/SPE Workshop: Next Generation Deepwater Facilities
December 12-14, 2016 | Hyatt Regency Houston/Galleria | Houston, TX

AIR LIQUIDE

Creative Oxygen

Improving Analytical Accuracy of Measurements

Agenda

- Consequences of Analytical Variability
- How to Minimize Analytical Variability
- Summary

Agenda

- ❑Consequences of Analytical Variability
- How to Minimize Analytical Variability
- Summary

Consequences of Variability in Analytical Results

- Extra time to reconcile data
- Additional costs
- Lost revenue
- Delay in deploying “appropriate corrections”
- Poor decision making
- Loss of credibility

Minimizing Error in Analytical Measurements

- Analytical uncertainty originates from 4 primary sources

Reference Material

Reference Standards
"True Value"

Analyzer Calibration

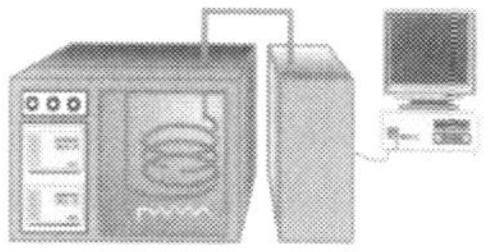

Analytical Instrument

Analyzer Precision

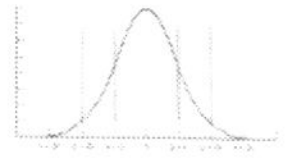

Result & uncertainty

Representative Sample

Understanding Error

- To obtain the total uncertainty of 2 or more quantities, each with their own uncertainty
- Measured values for the quantities X, Y & Z with uncertainties δX, δY & δZ
  - The total uncertainty (or accuracy), δU is

  - $\delta U = + \sqrt{(\delta X)^2 + (\delta Y)^2 + (\delta Z)^2}$

Understanding Error

- Minimize "Basic" Analysis Uncertainty Sources
  - Reference Material (U_S) A%
  - Analyzer Calibration (U_C) B%
  - Analyzer Precisions (U_P) C%
  - …… D%

Achieving an uncertainty less than that of the "reference material is not possible

- "Propagation of Error" Calculation
  - Accuracy (U_T) = $\pm \sqrt{(A)^2 + (B)^2 + (C)^2 + (D)^2}$
  - Accuracy (U_T) = $\pm$ x %

How Do I Get the Best Analytical Data?

- Minimize the Errors!

Agenda

- Consequences of Analytical Variability
- How to Minimize Analytical Variability
- Summary

1. Reference Material

- Why is having a reference important?

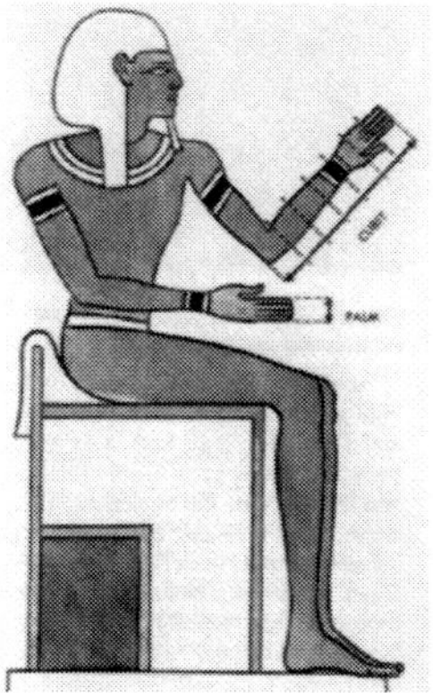

Egyptian cubit based on the length of the Pharaoh's forearm & hand

1 mile = 1000 Roman paces

Traceability Model

SI Units

↓

National Measurement Standards
National Metrology Institute

↓

Gas Industries

↓

Laboratory
or
Field measurement

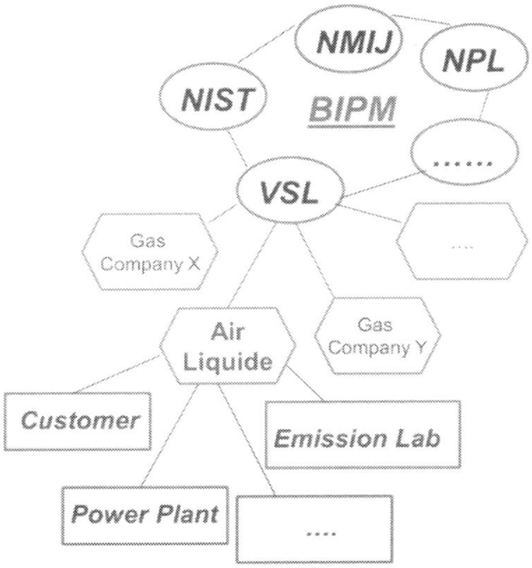

Reference Material

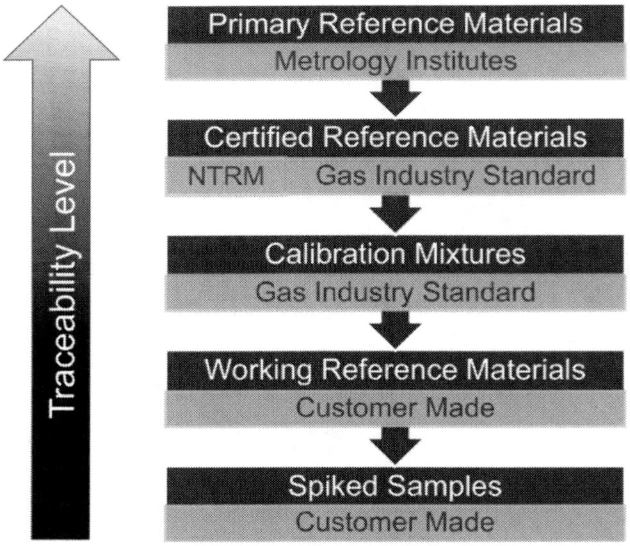

- Measurement accuracy is assessed by
  - Analyzing a sample with a known value and comparing it to a "known" value
  - Over multiple days with repeated analysis

Highest Accuracy Measurements Require Traceability

- Traceability of measurement requires an unbroken chain of comparisons to stated references
  - Each with a stated uncertainty
- Analysis of the mixture with an instrument directly calibrated with reference material from a recognized metrology organization such as NIST, VSL, NPL or KRISS, …
- When the analysis is performed with a NIST SRM (or NTRM), the resulting analyzed product is said to be "NIST traceable".

Measurement Accuracy

- The lower the uncertainty of the standard used to calibrate the sample
  - The higher the accuracy of your result will be

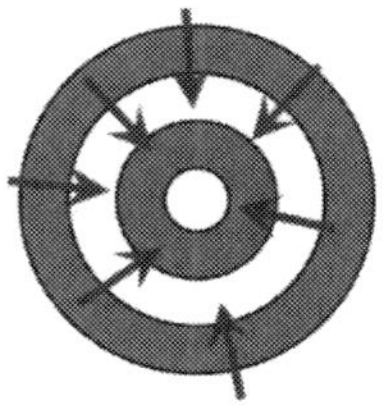

2. Obtaining a Representative Sample

- Is the sample taken representative of the sample stream?
  - Is a sample being taken at the correct frequency to identify a perturbation?
  - Does the sample accurately represent the population from which it is drawn?
- Are the compatible materials (with the sample stream) being used throughout?
  - Is the sample reacting with the transfer tubing?
  - Is the analyte of interest being adsorbed?

Material Compatibility

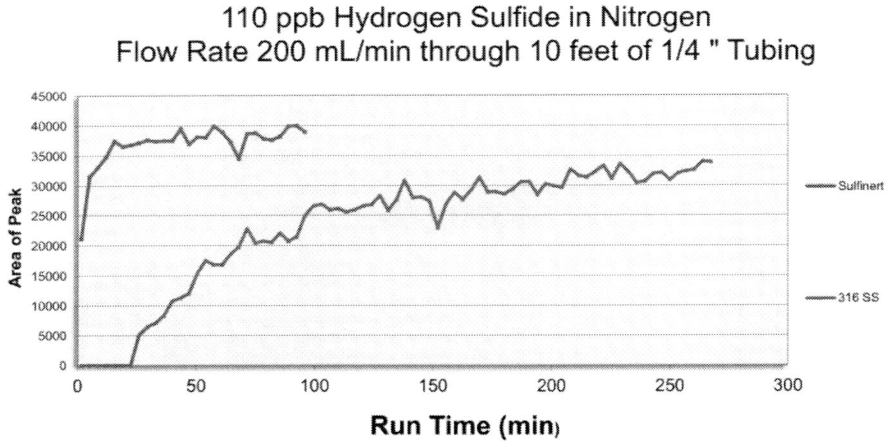

Biela, B., Moore, R., Benesch, R., Talbert, B., Jacksier, T.: "The Do's and Don'ts in the Analysis of Sulfur for Polyolefin Producers", Presented at the Gulf Coast Conference, Galveston Island, TX, Oct 2003

Material Compatibility

- GC inlet system: Sample loop materials

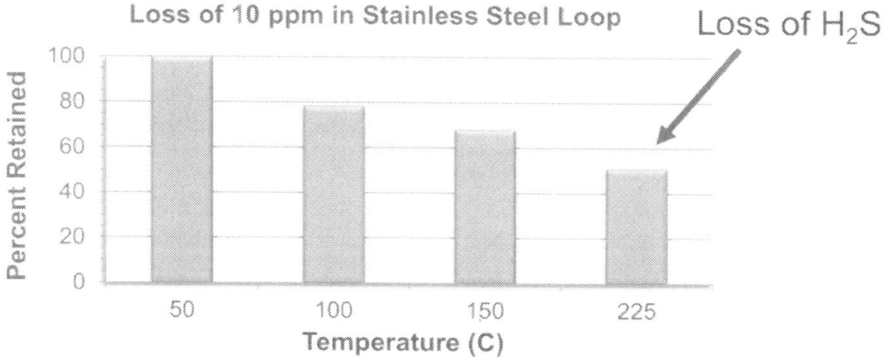

Silcosteel™ - no noticeable effect 50 °C to 225 °C (not illustrated in this figure)
Stainless Steel – considerable loss of signal

Haouchine, M., Benesch, R., Talbert, B., Jacksier, T., Biela, B., Moore.: "Monitor Sulfur Levels in your Polyolefin Feedstock", Hydrocarbon Processing, Oct (125) 2007.

3. Analyzer Calibration

- An essential part of the measurement procedure
- Establishes the relationship between instrument response and the accepted value (e.g. the concentration of the analyte present)

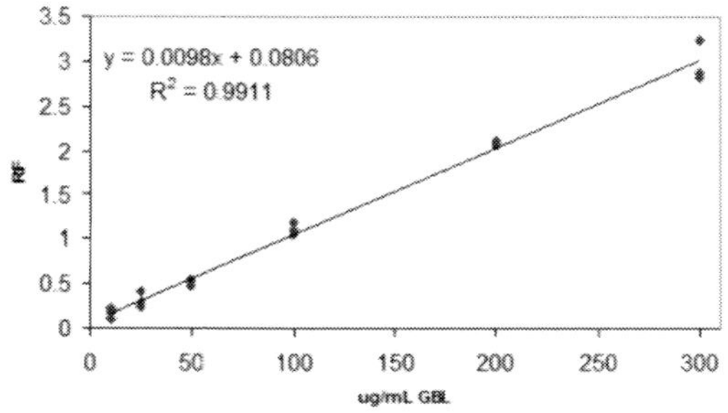

Principles of Calibration

Calibration using several concentrations

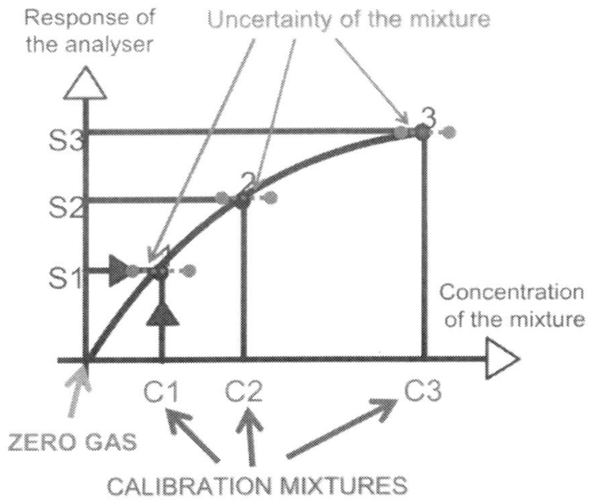

1) Check the zero of the analyser using a pure gas that does not contain the analyzed component (zero gas).
2) Introduce several mixtures of the component to analyze (calibration mixtures). Monitor the signal for each mixture.
3) Calculate the response curve of the analyser (its trueness depends on the uncertainty of the mixtures).

Principle of Calibration

Use of the calibration curve

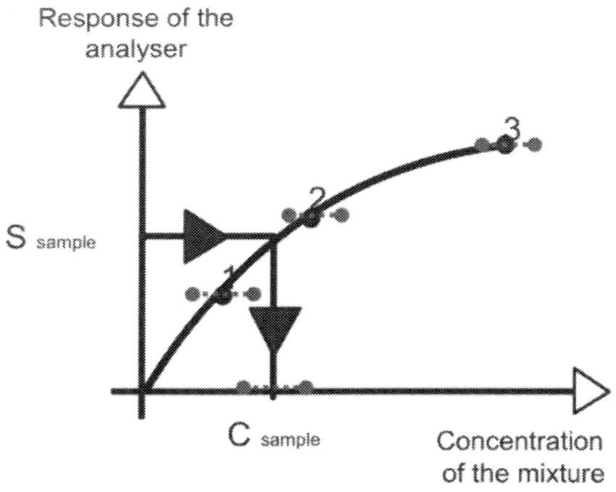

1) Introduce the sample to analyse in the analyser.

2) Monitor the signal obtained S_{sample}

3) Use the response curve of the analyser to determine the concentration of the sample C_{sample}

The uncertainty on the concentration C_{sample} depends on the trueness of the response curve.

Principle of Calibration

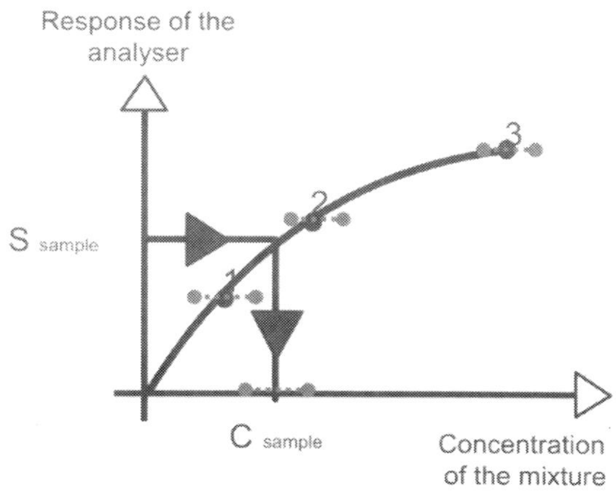

- The calibration range should be established so that the majority of the sample concentrations fall towards the centre of the range.

4. Analyzer Precision

- Recall the difference between accuracy & precision

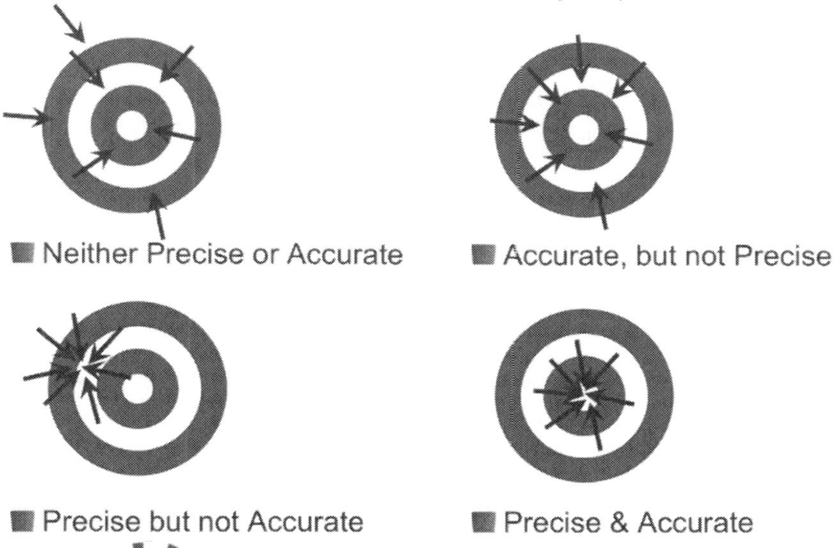

Analyzer Precision

- Precision describes the closeness with which multiple analyses of a given sample agree with each other, and is sometimes referred to as reproducibility.
  - Note that precision is concentration dependant
  - **Values should be calculated at low, mid and upper ranges**
  - The higher the number of samples used to validate the precision, the higher confidence in the actual measurement
    - Typically 5 or 7

Agenda

- Consequences of Analytical Variability
- How to Minimize Analytical Variability
- Summary

Putting it All Together

- To obtain the highest accuracy of measurement
  - Identify the required accuracy & precision for your particular analysis
  - Obtain calibration standards which satisfy these requirements
  - Validate analytical methodology
  - Experimentally validate sampling methods are representative and provide representative samples

THANK YOU FOR YOUR ATTENTION!

www.airliquide.com

Long Distance Subsea Oil Tiebacks – Considerations from an Operators Perspective

STEVEN JOHNSON
SUBSEA PROJECT MANAGER,
CHEVRON PROJECT RESOURCES COMPANY

AIChE The Global Home of Chemical Engineers
Joint AIChE/SPE Workshop: Next Generation Deepwater Facilities
December 12-14, 2016 | Hyatt Regency Houston/Galleria | Houston, TX
SPE International Society of Petroleum Engineers

DISCUSSION OVERVIEW

Key Discussion Areas

Project Opportunity

Team selection & alignment

Uncertainties in the Design Basis

Technology opportunities

Oil, & gas handling

Subsea components

Opportunity summary

AIChE The Global Home of Chemical Engineers
Joint AIChE/SPE Workshop: Next Generation Deepwater Facilities
December 12-14, 2016 | Hyatt Regency Houston/Galleria | Houston, TX
SPE International Society of Petroleum Engineers

PROJECT OPPORTUNITY

To develop remote oil fields that are currently deemed uneconomic, or not technically practical using accepted, established technology, even after ongoing cost reduction opportunities have been implemented

TEAM SELECTION & ALIGNMENT

Team selection and development

- Recognize project may require advanced technology
- Assemble co-located multidiscipline team, able to rapidly assimilate new ideas
- Integrate subsurface, subsea, topsides & operations personnel in project team
- Allow teambuilding & training time to educate team into available technology
- Allow time to assimilate ideas and their integration into the production system

Keep Corporate Management, Partners & other Stakeholders informed of progress. The solution that evolves may require un-conventional technology which needs to be "sold" throughout organization.

UNCERTAINTIES IN THE DESIGN BASIS

Project Design Basis can take several years to collect & interpret

- Seabed & environmental surveys
- Exploration/ appraisal well drilling
- Fluid sampling, testing & analysis
- Subsurface analysis

Develop analogue information that is refined over time

Life of field solution, not just to 1st oil.

Selection of host facility

Design Basis uncertainties will, in turn lead to project cost, schedule and economic uncertainties

TECHNOLOGY OPPORTUNITIES

Need to explore technology opportunities

- What is available internally within your organization?
- What do your Partners have that can be brought to this project?
- What is available within Industry?
- What other technology is being developed that may be applied, for example through Joint Industry Projects (JIP's)?

Develop multiple alternative field development solutions, from well to topsides, incorporating technology opportunities

- What technology is enhancing?
- What technology is enabling?

Run field development economics on these alternative field development solutions to understand their value

AIChE The Global Home of Chemical Engineers
Joint AIChE/SPE Workshop: Next Generation Deepwater Facilities
December 12-14, 2016 | Hyatt Regency Houston/Galleria | Houston, TX
SPE International Society of Petroleum Engineers

TECHNOLOGY OPPORTUNITIES

Understand Technology Status & Risk, then Rank Opportunities:

- Group 1: Concepts that can use existing technology, just assembled differently to provide project solution
- Group 2: Minor technology enhancements provide an opportunity for improving the economy, safety, reliability... need to manage maturity/ uncertainty to get real benefits
- Group 3: New technology, with technology development required

How are reservoir definition uncertainties managed in parallel?

Is there time, money & engineering resources available to develop that technology?

OIL & GAS HANDLING SUBSEA FACILITIES

Recognize alternative boosting technologies are available to enhance/ enable fluid flow rate to host;

- Gas lift
- Pumping & compression
- Subsea separation & boosting

Alternative locations for this boosting include;

- Riser base
- In-riser
- Wellhead/ manifold
- Downhole

OIL & GAS HANDLING HOST FACILITIES

Which Host is selected?

- Is there sufficient ullage capacity?
- Is de-bottlenecking needed?

What additional electrical power may be required?

- Is there space & weight capacity for any required additional switchgear, transformers & adjustable speed drives?

How will the new subsea controls & chemical injection system be incorporated?

How will the additional risers & umbilicals/ power cables be incorporated?

SUBSEA COMPONENTS

Trees, Manifolds & Controls

- Hydraulic or All-electric
- Start-up/ shutdown management
- Fiber optics for monitoring & communications

Flowlines

- Flow Assurance considerations
- Improved insulation & electric heating

Pumps, Compressors & Separators,

- Single phase or multiphase pumps
- Gas compressors
- 2 or 3 phase subsea separation

AIChE The Global Home of Chemical Engineers
Joint AIChE/SPE Workshop: Next Generation Deepwater Facilities
December 12-14, 2016 | Hyatt Regency Houston/Galleria | Houston, TX
SPE International Society of Petroleum Engineers

SUBSEA COMPONENTS

Electric Power

- High power cables
- Seabed transformers, switchgear & adjustable speed drives
- Electrical connectors

Water Injection

- From host platform
- Seabed water treatment
- Re-injection of produced water

Subsea chemical injection

- From host platform, or seabed storage & distribution

OPPORTUNITY SUMMARY

Assemble free thinking team to take this forward

Engage stakeholders throughout the process

Allow the Design Basis to mature as design develops

Recognize advanced subsea technology, that can enable long distance subsea oil tie-backs, is either commercially available, or close to being commercial

Take time to learn about this technology, what it is capable of, its value and how best to incorporate it.

Understand the associated risks & qualification requirements

Phasing – Today's Deepwater Strategy

STAFFORD MENARD

audubon
ENGINEERING SOLUTIONS

DISCUSSION TOPICS

Definition of Phasing

Small FPS Example

Project Timelines

Economic Evaluations

Summary- Pros and Cons

AIChE The Global Home of Chemical Engineers

Joint AIChE/SPE Workshop: Next Generation Deepwater Facilities
December 12-14, 2016 | Hyatt Regency Houston/Galleria | Houston, TX

SPE International Society of Petroleum Engineers

DEFINITION PER WEBSTER'S DICTIONARY

Phasing (1902)
To conduct or carry out by planned phases
To introduce in stages

WHAT DO WE MEAN BY PHASING

Development of a proven reservoir in a safe, reliable and economical manor by means of:

1) Utilizing a small pre-designed standard FPS (SFPS)
2) Checking-off on gates early (off critical path)
3) Drilling fewer appraisal wells
4) Drilling sufficient initial production wells to load SFPS
5) Drilling additional production wells as required
6) Using the production wells to evaluate full field development requirements
7) Adding an additional SFPS if required

SFPS EXAMPLE

- Production Capabilities
  - 40,000 BOPD
  - 60 MMSCFD
  - 15,000 BWPD
- Oil gravity – 27° to 37° API Sweet
- Meets BSEE &USCG requirements
- Designed for current central Gulf metocean conditions
- Quartering 38-40 POB
- Designed to company specific specifications/requirements, or third party existing FPS design

MAIN DECK GA

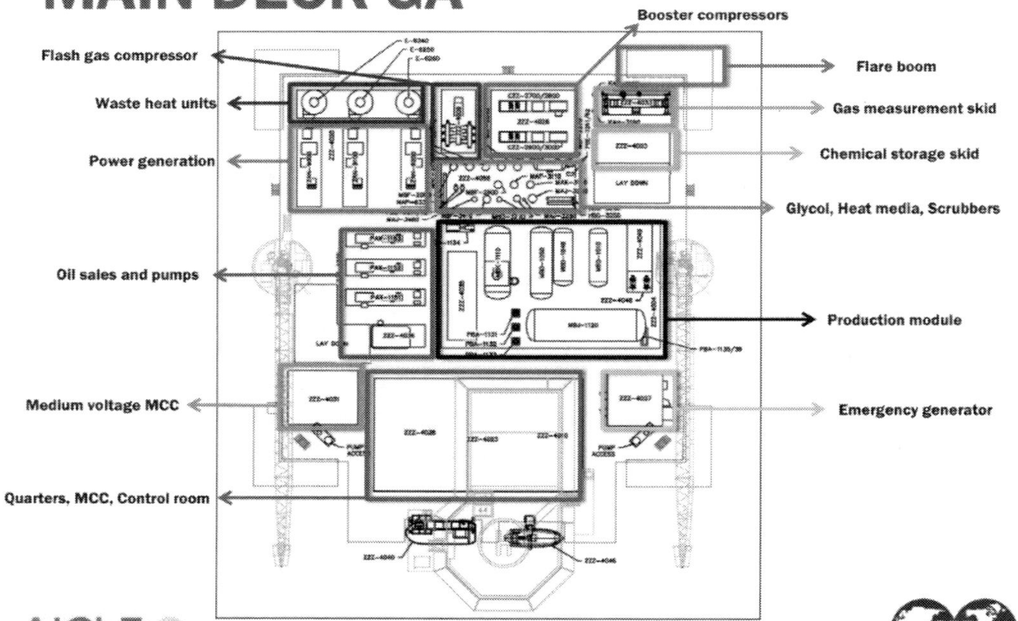

CELLAR DECK GA

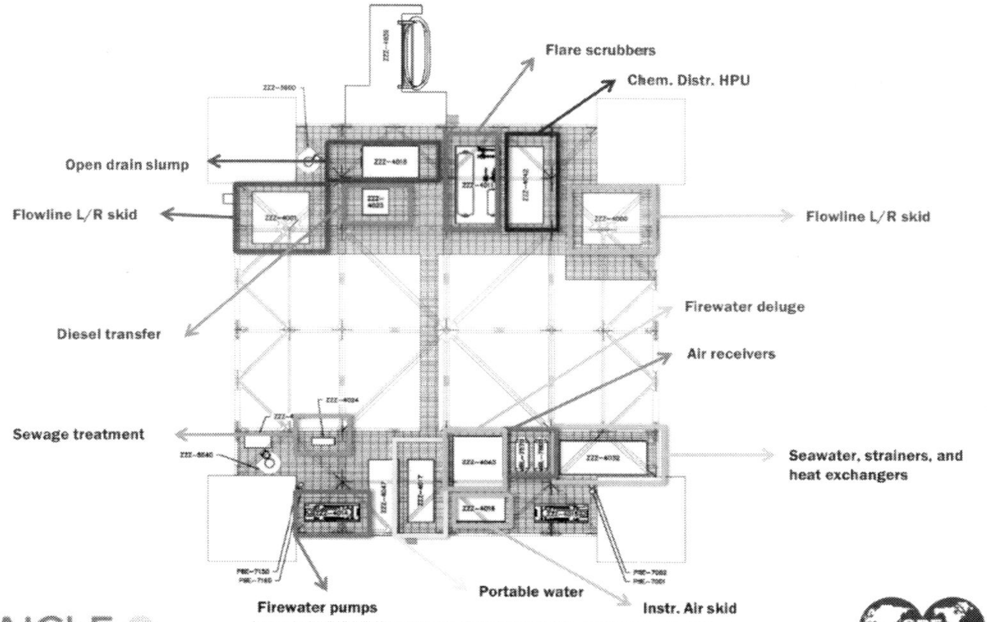

OPTIONS AND OTHER USES

Options to consider for the standard FPS

1) Seawater Injection System
2) Gas Lift
3) Mudline Pumping/ESPs
4) Other Process Treatments

Other uses

1) Early Production Systems (EPS)
2) Late field life replacement

TIME LINE – SFPS vs CONVENTIONAL FPS STRATEGY

Key Points:

Conventional Strategy – 1 FPS

Appraisal Wells – 5

Discovery to First Oil – 8 years

Phased Strategy – 2 SFPSs

Appraisal Wells – 2

Discovery to First Oil – 5 years

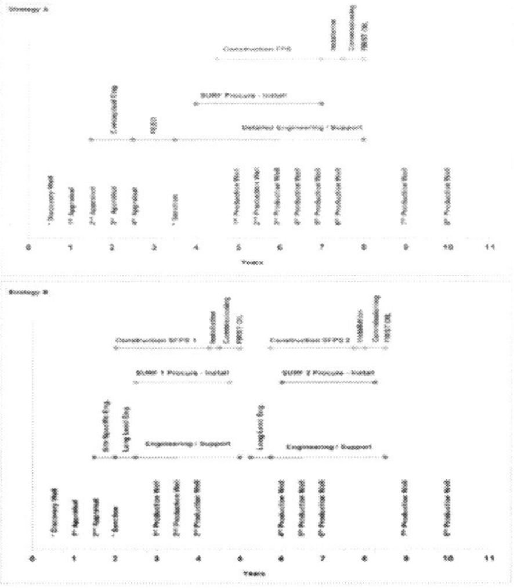

P-50 200 MMBOE FIELD

Assumptions:

Crude Price	**$50.00**
Quality Discount	**$0.0**
Tariffs	**$3.50**
Facility Opex	**$4.00**
Royalties	**$6.25**
Net Crude Price	**36.25**

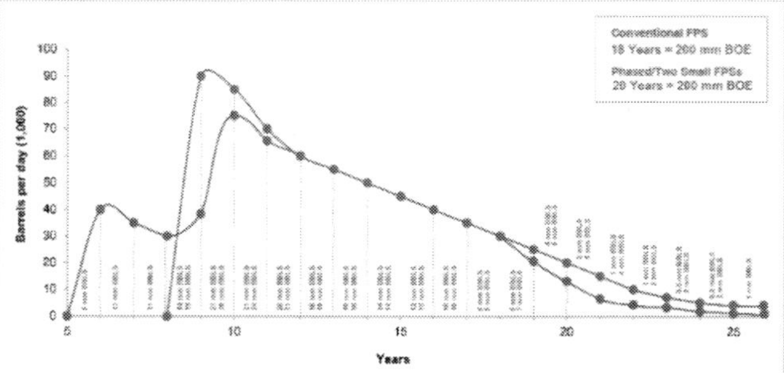

Crude Inflation	**2%**
Discount Rate	**11%**
Appraisal Wells	**$150MM**
Production Wells	**$180MM**
Cost Escalation	**2%**

	NPV	MIRR
Conventional:	**$395.5MM**	**8.3%**
Phased:	**$734.0MM**	**10.6%**

AIChE The Global Home of Chemical Engineers
Joint AIChE/SPE Workshop: Next Generation Deepwater Facilities
December 12-14, 2016 | Hyatt Regency Houston/Galleria | Houston, TX
SPE International Society of Petroleum Engineers

P-50 100 MMBOE FIELD

Assumptions:

Crude Price	**$50.00**
Quality Discount	**$0.0**
Tariffs	**$3.50**
Facility Opex	**$4.00**
Royalties	**<u>$6.25</u>**
Net Crude Price	**36.25**

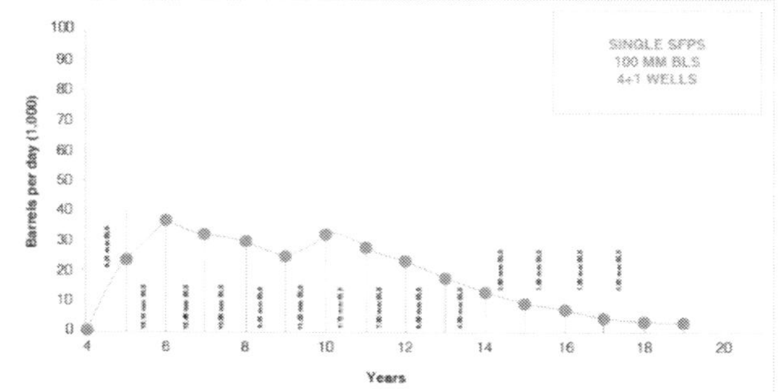

Crude Inflation	**2%**
Discount Rate	**11%**
Appraisal Wells	**$150MM**
Production Wells	**$180MM**
Cost Escalation	**2%**

NPV
$130 MM

MIRR
7.3%

PROS AND CONS OF PHASING/ STANDARDIZATION

CONS

- Complacency in:
  - Design
  - Project Execution
  - Operations
- Reduced Innovation
  - "This is the way we always do it"
- Does not provide 100% solution for all possible projects

PROS AND CONS OF PHASING/ STANDARDIZATION

PROS

- Reduced schedule to FIRST OIL
- Early CASH FLOW & improved NPV
- Lease secured through production
- STANDARD COMPONENTS (reduced cost)
- SIMPLIFIED execution plan & schedule
- REDUCED DEVELOPMENT RISK - Reduced investment
- Opportunity to phase (EXPAND) development if needed
- Expansion can be paid through production
- Potential reuse of one or both SFPS

AIChE
The Global Home of Chemical Engineers

Joint AIChE/SPE Workshop: Next Generation Deepwater Facilities
December 12-14, 2016 | Hyatt Regency Houston/Galleria | Houston, TX

Society of Petroleum Engineers

A History of Too Much Capacity and Too Much Complexity in the Deep Water Gulf of Mexico

JAMES R. DEAVER, P.E.

AIChE The Global Home of Chemical Engineers

Joint AIChE/SPE Workshop: Next Generation Deepwater Facilities
December 12-14, 2016 | Hyatt Regency Houston/Galleria | Houston, TX

Society of Petroleum Engineers

DISCLAIMER

- I Intend to Skewer a Few 'Sacred Cows'
- I Intend to Point at The 'Emperor's New Clothes'

But,

- I also Intend to Put Forward My Opinion of the Root Causes of Much of the Complexity
- I Intend to Propose Some Real World Solutions to Reduce Complexity, and Maybe Improve Project Performance

AIChE The Global Home of Chemical Engineers

Joint AIChE/SPE Workshop: Next Generation Deepwater Facilities
December 12-14, 2016 | Hyatt Regency Houston/Galleria | Houston, TX

Society of Petroleum Engineers

BENCHMARKING AND PERFORMANCE

In the last 15 years, deep water projects routinely:

- Significantly overrun their budgets
- Achieve first production significantly late
- Rarely achieve nameplate capacity, much less plateau
- Have very poor production efficiency (availability)
- Are fairly inflexible to 'off design point' operation

AIChE
The Global Home of Chemical Engineers

Joint AIChE/SPE Workshop: Next Generation Deepwater Facilities
December 12-14, 2016 | Hyatt Regency Houston/Galleria | Houston, TX

SPE International
Society of Petroleum Engineers

PRODUCTION PERFORMANCE

By the BSEE Definition, Deep Water is 1,000'

From 09-15-2016 BSEE website reports

- **19 operators, only four have more than 3 facilities, seven have one**
- **50 deep water facilities**
- **2016 total production yields 940,280 BOPD**
- **Total nameplate capacity is 3,790,500 BOPD**
- **Combined Industry Utilization is *24.8%***
- **Only one operator exceeded 50% (56.5%)**

AIChE
The Global Home of Chemical Engineers

Joint AIChE/SPE Workshop: Next Generation Deepwater Facilities
December 12-14, 2016 | Hyatt Regency Houston/Galleria | Houston, TX

SPE International
Society of Petroleum Engineers

PROJECT COMPLEXITY

- SPE currently has a lot of interest in the subject of project complexity.
- The industry likes to blame complexity for poor project performance.
- Unfortunately, our industry is extraordinarily good at solving technically complex problems. So this excuse is pretty weak.
- I put forward that the poor performance is self inflicted social complexity.

PRODUCTION FACILITY REQUIREMENTS

Fundamentally, production facility requirements in deep water are no different than onshore.

- Provide a safe work environment
- Produce wells
- Separate oil, gas, and water
- Treat each stream as necessary for disposition
- Measure each stream
- Dispose of each stream
- Treat and inject water for waterflood when required

And make a profit while doing so.

WHAT FACILITIES HAVE BECOME

- Showcases of cleverness
- Confusing in the name of flexibility
- Not understood by the people operating them
- Designed for an operating point, not a naturally declining envelope
- Customized because we know better than the manufacturers what features we need them to add to our special application
- Priced by the page of specification, or priced by the client's reputation

COMPLEXITY SOURCES

- The ability to generate unlimited numbers of pages in the 'paperless' society for a specification, and then (sort of) repeat in multiple other specifications
- The proliferation of specialists and specialist studies
- Specialists 'optimizing' at cross purposes or without regard for other specialist optimization
- The lack of a generalist in a role with sufficient authority
- The lack of any individual in a role with sufficient authority
- The blind belief in systems and procedures to guaranty quality rather than using them to support individual purpose

COMPLEXITY KILLERS

- Start the design with a clear definition of the problem you're trying to solve, in as few words as possible
- Simplicity is good; usually safer and more operable
- Limit studies to the ones you don't already know the answer
- Try to get by with as little as possible. System optimization doesn't matter as much as elimination does
- Plan for turndown and reduced pressure operation
- Plan for start-up and maintenance

AIChE
The Global Home of Chemical Engineers

Joint AIChE/SPE Workshop: Next Generation Deepwater Facilities
December 12-14, 2016 | Hyatt Regency Houston/Galleria | Houston, TX

SPE International
Society of Petroleum Engineers

COMPLEXITY KILLERS

- Debottlenecking is the most efficient money ever spent
- Don't confuse safety with convenience
- Recognize diseconomy of scale
- Remember that saving money is almost always more important than saving weight

AIChE
The Global Home of Chemical Engineers

Joint AIChE/SPE Workshop: Next Generation Deepwater Facilities
December 12-14, 2016 | Hyatt Regency Houston/Galleria | Houston, TX

SPE International
Society of Petroleum Engineers

THANK YOU

Contact info

James R. Deaver, P.E.

Engineering Advisor

Oil Field Development (OFD) Engineering

281-556-4991

james_deaver@ofdeng.com

AIChE The Global Home of Chemical Engineers

Joint AIChE/SPE Workshop: Next Generation Deepwater Facilities
December 12-14, 2016 | Hyatt Regency Houston/Galleria | Houston, TX

SPE International Society of Petroleum Engineers

GENESIS

Ashutosh Kak and Oscar Coronado
December 12-14, 2016

Life Cycle System Design Approach For Flow Assurance

Joint AIChE/SPE Workshop: Next Generation Deepwater Facilities

Hyatt Regency Houston/Galleria | Houston, TX

AIChE
The Global Home of Chemical Engineers

SPE International
Society of Petroleum Engineers

www.genesisoilandgas.com

GG-COM-TM-013-A, Rev. 2

Disclaimer

Information contained in this document is the work product of Genesis North America and embodies confidential and proprietary information of Genesis North America and/or its client. This document and all information herein are confidential, and may not be used, reproduced or distributed without prior authorization of Genesis North America. Genesis North America disclaims any and all liabilities for any changes or modifications to this document which are made by any non-Technip party.

Acknowledgement

- Support from Genesis Management
- Contributions from Dr. Weihong Meng, PC Chandrahasan, and Dr. Oscar Coronado

Intent

- How do we make it work in 40$/barrel

- I have few questions for you

Questions

- Conceptual throughputs: what's the basis?
- Design margins in engineering: can we be less conservative?
- Can we include operations feedback to improve operability and hence OPEX?

Create > Realize > Enhance

CONCEPTUAL THROUGHPUTS WHAT'S THE BASIS?

Production Forecast

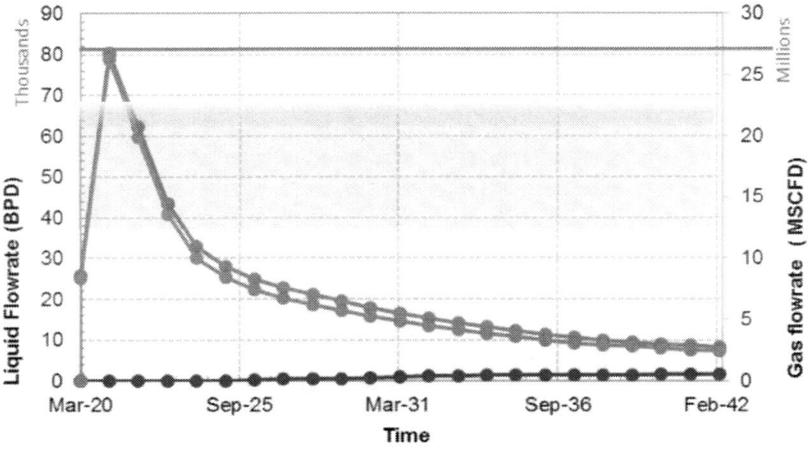

Should we design the facility to accommodate the maximum production rate?

Optimize facility design capacity by considering CAPEX, OPEX, and maximum recovery

Higher throughout design approach may lead to quicker payoffs, if the economics is run only for few years. However; when the life-cycle economics is considered, the cons will far out-weight the pros.

Historical Data

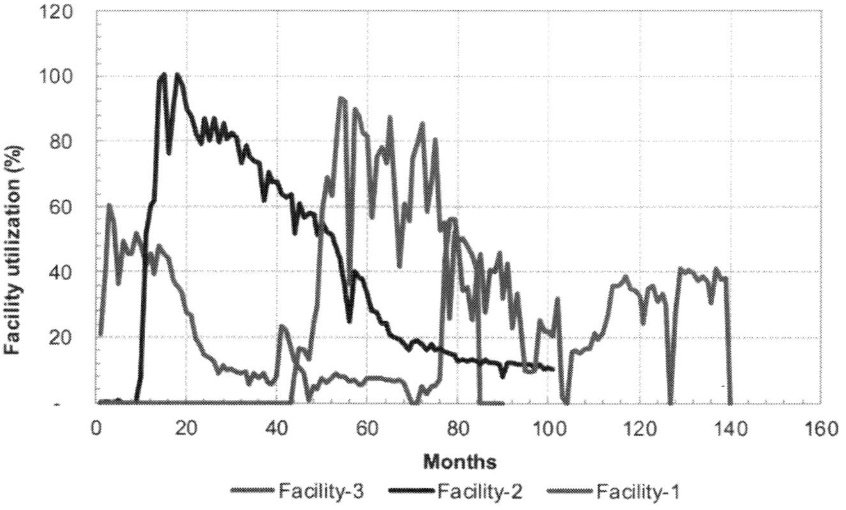

For simplifications purpose this data assumes all months have equal duration of 30 days

Create > Realize > Enhance

Historical Data

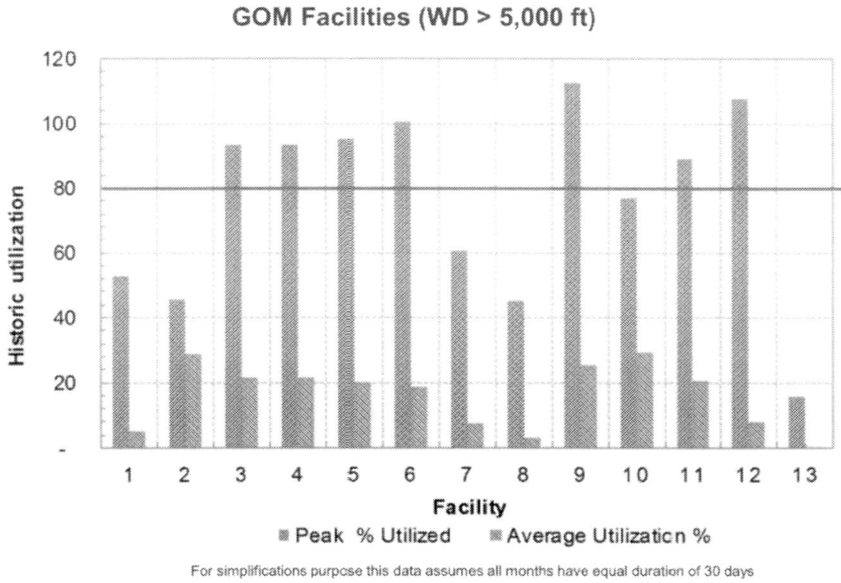

Can the maximum facility design capacity be reduced?

Lower CAPEX

Create > Realize > Enhance

Operating Envelope

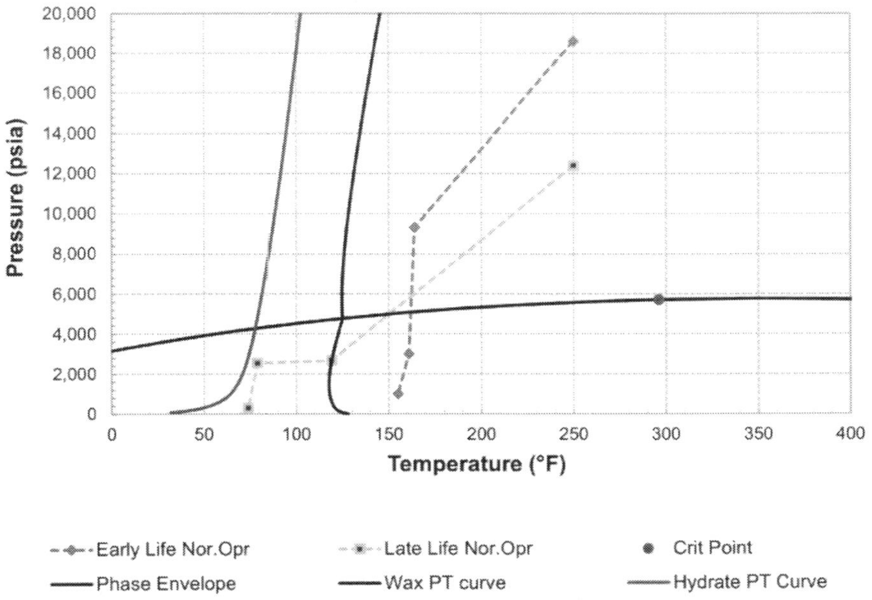

Impact on Production Pipelines

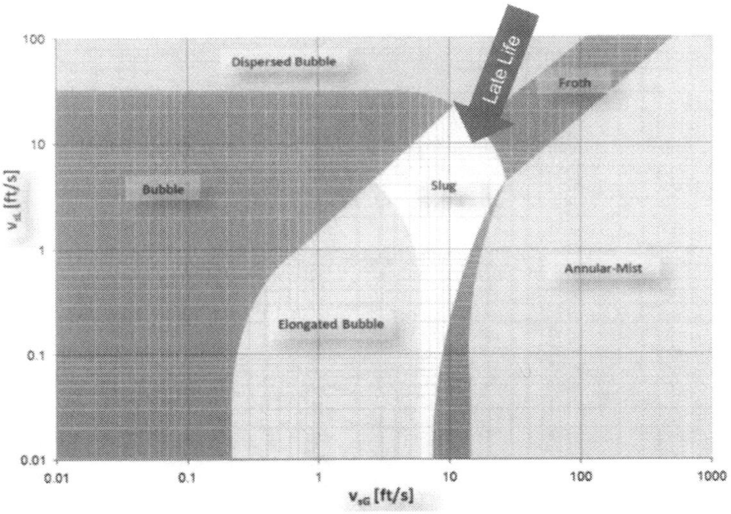

- Reduction of maximum subsea choke pressure differential in early life
- Reduce heat loss
- Extend field life by higher flow velocity, and less slugging
- Save material and installation cost
- Reduce drilling and completion cost

Image Courtesy: fekete.com

Issues

- Separator performance
  - Slugging causes shutdown
  - Low temperatures due to expansion cooling
- Gas compression and availability
  - Gas shortage for gas lift systems
- Thermal Management
  - Hydrates
  - Wax
- Higher well back- pressure
  - Slower recovery rate
  - Lower ultimate recovery

GENESIS

DESIGN MARGINS IN ENGINEERING CAN WE BE LESS CONSERVATIVE?

Create > Realize > Enhance

www.genesisoilandgas.com

Create > Realize > Enhance

Insulation Design for Flowlines

- Uncertainties in Pre-FEED and FEED
  - ➢ Self-burial of pipe
  - ➢ Well completions
    - Final tubing configuration
    - Thermal performance
    - Drawdown limitations
  - ➢ Material properties

Example: Effects of Pipe Burial on the OHTC

- There are models that can predict the self burial of the pipeline
  - Typical burial is between 30 and 50 %
  - Example show impact of complete 100% burial

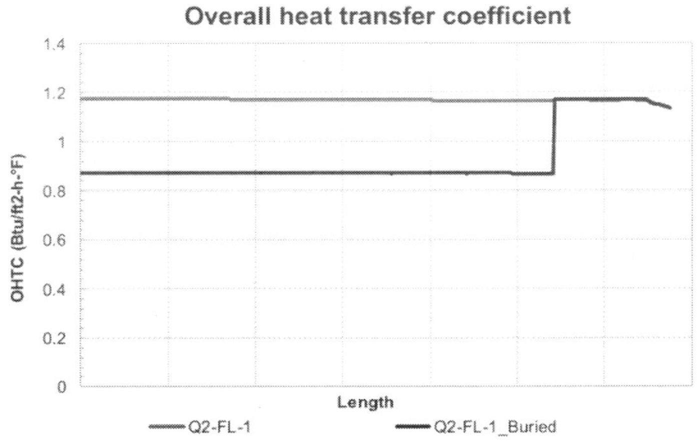

Create Realize Enhance

Example: Effects of Pipe Burial on the OHTC

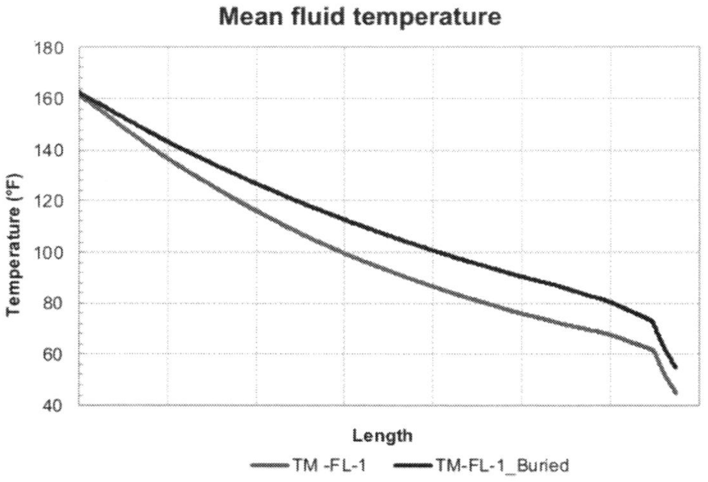

Confidential – Do Not Disclose Without Authorization © Copyright Genesis North America - All Rights Reserved

Example: Effects of Pipe Burial on the OHTC

Create > Realize > Enhance

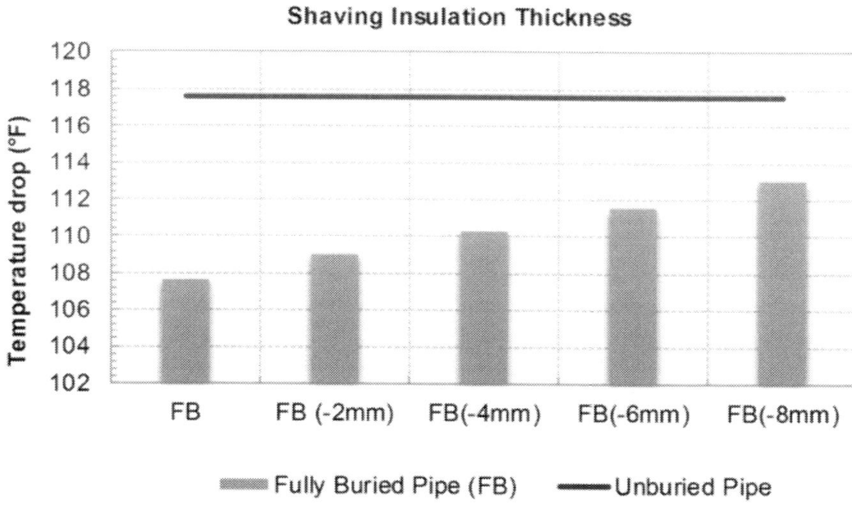

Insulation Cost

Create > Realize > Enhance

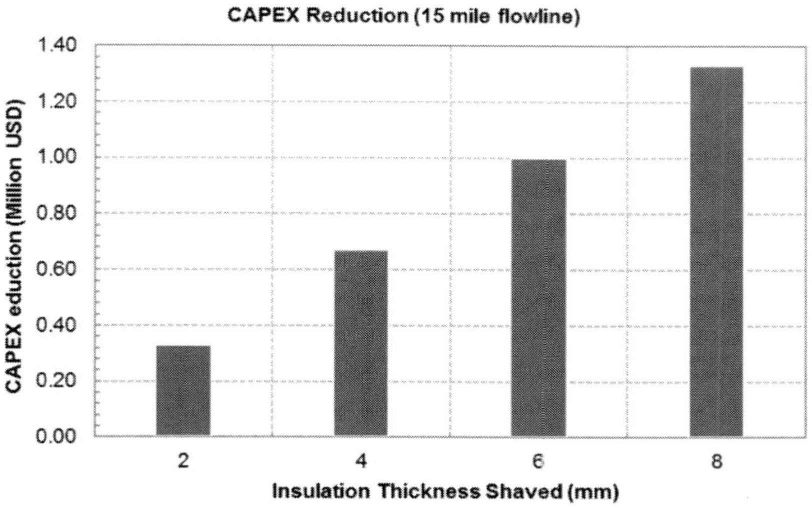

GENESIS

OPERATIONS FEEDBACK TO IMPROVE OPERABILITY AND HENCE OPEX ?

Create > Realize > Enhance

www.genesisoilandgas.com

Create > Realize > Enhance

Issues

- Inappropriate operating strategies
- Uncertainties in chemical properties
- Changing operating envelopes and subsea distribution
- Vacuum and related control issues
  - Run away injection
  - Overdosing / uneven dosing
  - Excess consumption

Example: Chemical Injection System Vacuum Issues

- When could it happen?
  - High GoR produced fluids
  - Later in life, because of declining operating pressures
  - Oversize chemical injection tubing

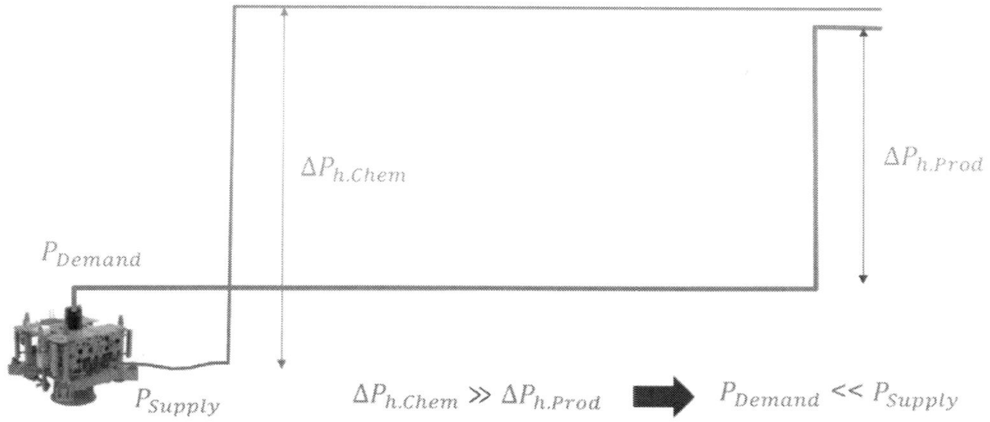

$\Delta P_{h.Chem} \gg \Delta P_{h.Prod}$ ➡ $P_{Demand} << P_{Supply}$

$(\Delta P_{h.} = \rho . g . h)$ *Vacuum Issues*

Example: Chemical Injection System Vacuum Issues

- Provision for Solutions in early stages
  - High pressure drop check valves installed before the injection point
  - Back pressure regulators (SkoFlow, Huntington)
  - Tubing coil
  - Nitrogen cap

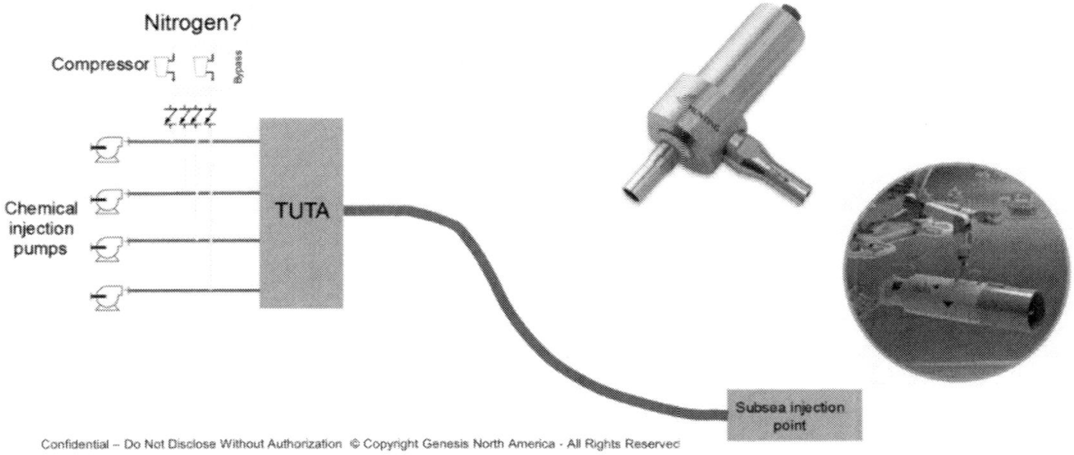

THANK YOU

Contact

Ashutosh.kak@genesisoilandgas.com

Value for Life PASSION FOR QUALITY

Proximal Chemical Storage and Injection – An Aid to Early Production Systems and Long Term Design?

PAUL STEAD – BAKER HUGHES UPSTREAM CHEMICALS

WHAT IS PROXIMAL STORAGE & INJECTION?

- **For subsea & tieback wells, seabed chemical storage and injection facilities at the well site**
  - No subsea chemical umbilical supply
  - May or may not be dependent on power supply from remote host
  - Potentially enabling technology for long reach tiebacks

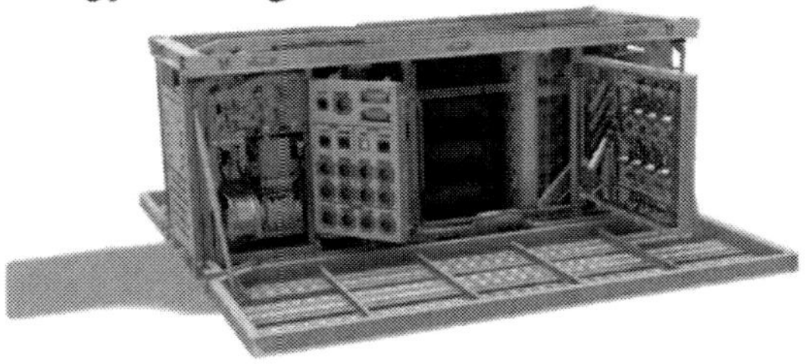

ADVANTAGES, DISADVANTAGES & CHALLENGES

- **Advantages**
  - Reduced capex
  - Broader range of chemical treatment options
    - Less restrictions on viscosity, allows for use of more active production chemicals
  - Increased flexibility
    - Potentially allows modular approach to treatment of flow assurance challenges during early production
    - Flow assurance uncertainty
  - Improved long term design

- **Disadvantages & challenges**
  - More complex supply chain
  - Some reduction in flexibility
    - More challenging to switch production chemicals
  - Power supply
    - May still require power from remote host as current battery technologies limiting stand alone use
  - Reliability?

AIChE
The Global Home of Chemical Engineers

Joint AIChE/SPE Workshop: Next Generation Deepwater Facilities
December 12-14, 2016 | Hyatt Regency Houston/Galleria | Houston, TX

SPE International
Society of Petroleum Engineers

BENEFITS FOR EARLY PRODUCTION & LONG TERM DESIGN

- **Temporary installment**
  - Reduced upfront Capex for EPF
- **Better design decisions & long term system efficacy**
  - Allows for confirmation of need and type of flow assurance chemistries prior to full field development
- **Could simplify contractual issues with third party operated host facilities**
  - No topside host chemical injection footprint or storage requirements

AIChE
The Global Home of Chemical Engineers

Joint AIChE/SPE Workshop: Next Generation Deepwater Facilities
December 12-14, 2016 | Hyatt Regency Houston/Galleria | Houston, TX

SPE International
Society of Petroleum Engineers

CURRENT STATE, DESIGN & READINESS

- **Autonomous Flooding & Testing Apparatus (AFTA)**
  - Designed for flooding & testing subsea lines
  - Built and ready to be deployed
  - System rated to 4,000m (~13,000 ft) water depth
  - Currently uses venturi system for dosing
    - Would require development of suitable small injection pumps
  - Scale, paraffin, corrosion, & asphaltene inhibitors typically injected at 20 -1000 ppm, so skid could give a viable operating duration of 3 to 12 months

CURRENT STATE, DESIGN & READINESS

- **Autonomous Flooding & Testing Apparatus (AFTA)**
  - Current battery technology would effectively limit operation to 3 – 6 months
  - AA type LDHI at 1 to 2% injection rate would only be sustainable for up to 10 days
  - Methanol or MEG injection at up to 50% injection rates would not be practical for more than a day without recharging

CURRENT STATE, DESIGN & READINESS

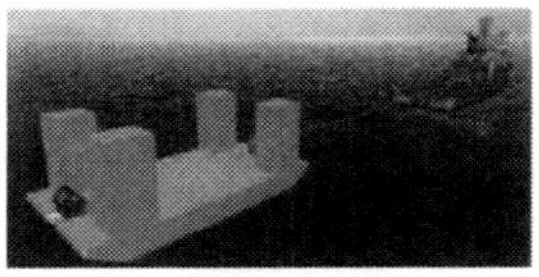

- **Safe Marine Transfer (SMT) system**
  - Two designs
    - One for multi-well, large developments
      - Chemical storage in hull (3 x 1100 bbls)
      - Could provide > 12 months supply of AA LDHI or 2 to 6 months supply of methanol/MEG
    - One for single well, small developments
      - Chemical storage for 200 bbls (or multiples thereof)
      - Could provide a number of months operation for scale, paraffin, corrosion, & asphaltene inhibitors

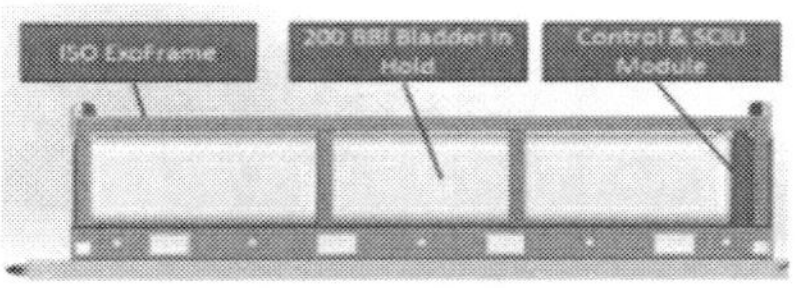

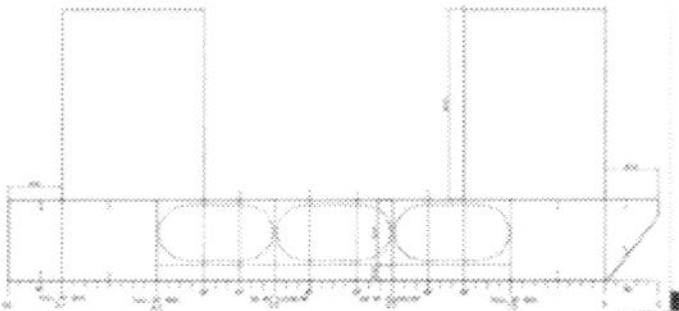

CURRENT STATE, DESIGN & READINESS

- **Safe Marine Transfer (SMT) system**
  - Components designed & fabricated, extensive testing conducted and reviewed with US regulators with approval pathway understood

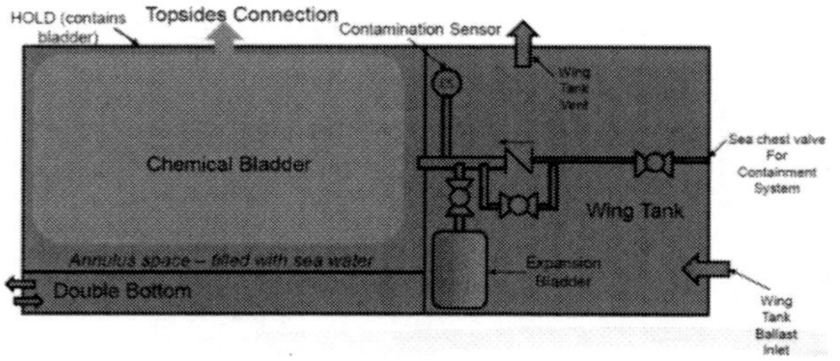

Pump/Battery Manifold

- **Other systems**
  - Various operators looking at similar systems

SUMMARY

- **Multiple system designs available for subsea chemical storage & injection**
- **Good potential for use in conjunction with early production facilities (EPF) to aid full field development design and long term system efficacy**
- **Potential for improving economic attractiveness of long reach tiebacks**
- **Systems still require some development and investment**
  - Improved battery technologies
  - Pumps
  - Reliability?

ACKNOWLEDGEMENTS

- **Thanks to the following for their input and for allowing the use of materials and information**
  - Art Schroeder, Jim Chitwood and all the folks at Safe Marine Transfer
  - Andy Barden, Baker Hughes Process & Pipeline Services

AIChE The Global Home of Chemical Engineers
Joint AIChE/SPE Workshop: Next Generation Deepwater Facilities
December 12-14, 2016 | Hyatt Regency Houston/Galleria | Houston, TX
SPE International Society of Petroleum Engineers

Questions?

AIChE
The Global Home of Chemical Engineers

Joint AIChE/SPE Workshop: Next Generation Deepwater Facilities
December 12-14, 2016 | Hyatt Regency Houston/Galleria | Houston, TX

SPE International
Society of Petroleum Engineers

Understanding the Production Chemistry and Other Flow Assurance Issues with Ambiguous Basis of Design Data

GRAEME SMITH / WALLY GEORGIE
MAXOIL PROCESS CONSULTANCY

maxoil Process Consultancy

AIChE The Global Home of Chemical Engineers

Joint AIChE/SPE Workshop: Next Generation Deepwater Facilities
December 12-14, 2016 | Hyatt Regency Houston/Galleria | Houston, TX

Society of Petroleum Engineers

Contents

- KEY DATA AND CHALLENGES
- DATA SOURCES - SAMPLING AND ANALYSIS
- CASE STUDY 1
- CASE STUDY 2
- CASE STUDY 3
- CONCLUSIONS

Key Data and Challenges

➢ **OIL CHARACTERIZATION**

- ➢ Asphaltenes – pressurized samples are preferred, normally contaminated by the drilling mud
- ➢ Naphthenates – most common is TAN value, but often misleading and requires additional sampling and costly testing.
- ➢ Paraffins – pressurized sampling downhole is the best option. Surface sampling is cheaper but often not representative

AIChE The Global Home of Chemical Engineers

Joint AIChE/SPE Workshop: Next Generation Deepwater Facilities
December 12-14, 2016 | Hyatt Regency Houston/Galleria | Houston, TX

Society of Petroleum Engineers

Key Data and Challenges

➢ **WATER CHEMISTRY**

- ➢ Getting suitable water sample can be costly
- ➢ In many cases water samples are not possible to collect
- ➢ Produced water samples are required for a variety of reasons
- ➢ Scaling Potential - Topsides/Subsurface/Subsea – often not representative. During initial sampling fluids are mainly oil with little water, and often contaminated by the drilling fluid
- ➢ Wrong Ca, Ba, Strontium is common and can mislead the design of the facilities, especially if there is water injection as part of the development

AIChE The Global Home of Chemical Engineers

Joint AIChE/SPE Workshop: Next Generation Deepwater Facilities
December 12-14, 2016 | Hyatt Regency Houston/Galleria | Houston, TX

Society of Petroleum Engineers

Key Data and Challenges

- **GAS COMPOSITIONS**
  - PVT sampling is good for HC gas composition
  - Acid gases - CO_2, H_2S
  - Difficult getting representative data – losses during sampling
- **CONTAMINANTS**
  - Mercury and other heavy metals – often an issue for gas condensate and LNG
  - Mercaptans – must be removed if at high concentrations
  - Diamondoids – significant challenge in certain areas of GoM have . Must be well understood and controlled. Can be dissolved by diesel, but requires large volumes

AIChE
The Global Home of Chemical Engineers

Joint AIChE/SPE Workshop: Next Generation Deepwater Facilities
December 12-14, 2016 | Hyatt Regency Houston/Galleria | Houston, TX

SPE International
Society of Petroleum Engineers

Data Sources - Sampling and Analysis

- **SAMPLES**
  - DST or pressurized sampling downhole – both have associated challenges
  - During Drilling Phase
    - Normally Contaminated
    - Small Volumes
    - Time constrained and expensive
  - All new wells/fluids should be tested
    - Update design data early
- **LAB TESTING**
  - Preservation and transport must be correct or results may not be representative
  - Aging of sample can often lead to quality issues
  - Measurement techniques vary between labs
  - Must be clear on whether data is tested for, or not available

AIChE
The Global Home of Chemical Engineers

Joint AIChE/SPE Workshop: Next Generation Deepwater Facilities
December 12-14, 2016 | Hyatt Regency Houston/Galleria | Houston, TX

SPE International
Society of Petroleum Engineers

Data Sources - Sampling and Analysis

- **OVERALL, THE INPUT TO THE DEVELOPMENT IS DEPENDENT ON THE QUALITY OF THE SAMPLING AND THE ANALYSIS**
  - Can vary from region to region and company to company
  - Will be used for design criteria
    - Separation
    - Flow assurance
    - Treatment strategies
    - Facilities required
    - Subsurface/subsea/topsides

AIChE
The Global Home of Chemical Engineers

Joint AIChE/SPE Workshop: Next Generation Deepwater Facilities
December 12-14, 2016 | Hyatt Regency Houston/Galleria | Houston, TX

SPE International
Society of Petroleum Engineers

Case Study 1

- **SOUTH EAST ASIA - FPSO**
- **PROBLEM**
  - Poor/Ambiguous Water Characterization
  - Used analogue data – from wrong field and old data
  - Underestimated Sand Production
  - Mercury in produced fluids but not found in original testing
- **WHAT HAPPENED?**

  Scaling
  - Wrong water chemistry
    - Lab testing reported ambiguous results for key scaling ions
    - Predicted topsides scale risk only
    - Scale inhibitor injection system provided for topsides only - not available subsurface or subsea
  - Scaling in wells, subsea flowlines and topsides

AIChE
The Global Home of Chemical Engineers

Joint AIChE/SPE Workshop: Next Generation Deepwater Facilities
December 12-14, 2016 | Hyatt Regency Houston/Galleria | Houston, TX

SPE International
Society of Petroleum Engineers

Case Study 1

- **WHAT HAPPENED?**
  - **Sand and Mercury Contamination**
  - High sand production early in production life
    - Most cases are guessed qualitatively
    - Removal facilities inadequate
  - Mercury produced with reservoir fluids
    - Contaminated process equipment
    - Contaminated sand
    - Sand removal and cleaning not designed for mercury content
    - Mercury exposure risks for vessel entry and maintenance tasks including sand removal

Case Study 1

- **END RESULT**
  - New subsea Scale Inhibitor injection lines installed
  - 500+ valves to be replaced (topsides)
  - New procedures required to account for mercury exposure – extended maintenance times
- **LESSON LEARNED**
  - Avoid ambiguous reporting of water chemistry
  - Ensure suitable analogues are used
  - Where feasible and data is unclear, be more conservative in the design

AIChE The Global Home of Chemical Engineers
Joint AIChE/SPE Workshop: Next Generation Deepwater Facilities
December 12-14, 2016 | Hyatt Regency Houston/Galleria | Houston, TX
SPE International Society of Petroleum Engineers

Case Study 2

- **NORTH SEA DEVELOPMENT**
- **PROBLEM**
  - Design Data (from exploration phase)
    - Low reported H_2S content
    - High mercaptan content
  - Actual Data (from early production phase)
    - High H_2S content
    - Low mercaptan content
- **END RESULT**
  - Exceeded allowable export H_2S content in the oil phase
  - Too high to chemically treat
  - New H_2S stripping (oil phase) equipment required
  - New bridged linked facility installed to host the new H2S stripping equipment
  - High CAPEX Mercaptan removal facilities installed but not required

AIChE The Global Home of Chemical Engineers

Joint AIChE/SPE Workshop: Next Generation Deepwater Facilities
December 12-14, 2016 | Hyatt Regency Houston/Galleria | Houston, TX

SPE International Society of Petroleum Engineers

Case Study 3

- **AFRICA**
- **PROBLEM**
  - Scaling of wells and flowlines with Barium Sulfate scale
  - No downhole scale inhibitor facilities installed
  - No provision for injection LSSW as thought to be no significant Barium level in formation water
- **WHAT WENT WRONG?**
  - Analysis of original water samples collected during drilling stated no Barium present
  - Lab had actually found high Barium, but thought it was an error in the analysis and reported no Barium
  - The design was based on no potential for Barium scaling
  - When production started, water re-analyzed and again showed high Barium
  - Remediation required expensive scale squeezes

AIChE The Global Home of Chemical Engineers

Joint AIChE/SPE Workshop: Next Generation Deepwater Facilities
December 12-14, 2016 | Hyatt Regency Houston/Galleria | Houston, TX

SPE International Society of Petroleum Engineers

Conclusions

- ACCURATE AND RELIABLE DATA IS KEY
- FLUID SAMPLES – BE PREPARED TO OBTAIN MORE TO ENSURE ALL TESTING (AND RE-TESTING) THAT IS NEEDED CAN BE PERFORMED
- ENSURE THERE ARE NO AMBIGUITIES IN THE REPORTED DATA
- PROVIDE INTERFACE BETWEEN DIFFERENT DISCIPLINES AND PROJECTS TO ENSURE CORRECT INTERPRETATION OF DATA
- WORK CLOSELY WITH THE LABS TO INVESTIGATE ANY ANOMALIES
- ENSURE ALL FLOW ASSURANCE AND CHEMISTRY ISSUES ARE PROPERLY ADDRESSED
  - Avoid unnecessary subsurface/subsea injection, but
  - Where necessary ensure there are sufficient facilities provided

Questions?

AIChE The Global Home of Chemical Engineers
Joint AIChE/SPE Workshop: Next Generation Deepwater Facilities
December 12-14, 2016 | Hyatt Regency Houston/Galleria | Houston, TX
SPE International Society of Petroleum Engineers

Contact Information

GRAEME SMITH

MAXOIL SOLUTIONS

maxoil
Process Consultancy

GRSMITH@MAXOILSOLUTIONS.COM

832 230 8246 / 713 471 8332

INTECSEA
WorleyParsons Group

Joint AIChE/SPE Workshop:
Next Generation Deepwater Facilities
December 12-14, 2016 | Houston, TX

Riser Friendly Floater Reducing Field Development CAPEX and OPEX

Alaa Mansour, Ph.D.
Marine Engineering Manager

Slide 2

Outlines

- Field Development Challenges
- Floaters Types and Limitations
- Riser Types and Limitations
- New Generation Riser Friendly Floater
- Case Studies: Reduced Field Development Costs
  - A Dry-Tree Application in South China Sea
  - A Wet-tree Application in West of Shetland
- Conclusions

Slide 3

Field Development Challenges

- **Lack of Infrastructures/Remoteness** ➡ **FPSO or FPU+FSO/FPSO**
  - Production storage is required
- **Requirement for DVA to wells** ➡ **TTR feasibility or Separate DTU**
  - Enhanced reservoir recovery
  - Mitigates flow assurance issues
  - Enables workover and completion from platform
  - Reduce drilling cost - **a case study (by KBR) shows over 1 Billion USD saving over wet tree**
- **Water depth**
- **Persistent swells**
- **Harsh environment**
- **Large number/diameter risers**
- **High dynamic floaters**
- **Riser design pressure**

➡ **Riser feasibility**

Slide 4

Floaters Types and Limitations

Slide 5

Floaters Types and Limitations

The Traffic Light Key:

- ● No Issues - Business as usual
- ● Issues that can be resolved with further engineering
- ● Unproven or Major issues that may not be resolved
- ● Major issues or show stopper

Slide 6

Floaters Types and Limitations

	FPSO	Semi	Spar	TLP

Market Share	70%	30%		
Production Storage Capability	●	●	●	●
Need turret in harsh environemnt	●	●	●	●
Suitability for very large payloads/ Deck Area	●	●	●	●
Quayside Integration	●	●	●	●
Suitability for ultra-deepwater application WD > 6,000 ft (1,829 m)	●	●	●	●
Motion Response	●	●	●	●
Suitability for Dry Tree Applications (TTRs)	●	●	●	●
Suitability for Steel Catendary Risers (SCRs) or Steel Lazy Wave Risers (SLWRs)	●	●	●	●

Slide 7

Risers Types and Limitations

Top Tensioned Riser

Total Stroke <8.5m (28') in 100yr
Used in water depth of 2,450 m
Diameter ~21" OD
High pressure 12 ksi

Courtesy of http://www.axiom.us.com/projects/themomentumisbuilding/prod_motion_compensation_tensioning_systems.html

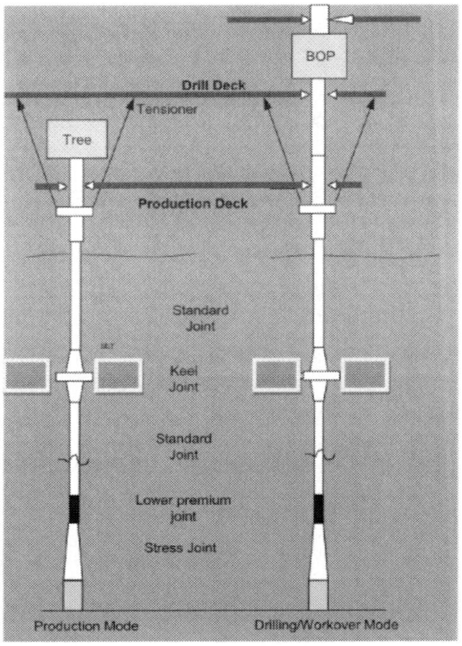

Slide 8

Risers Types and Limitations

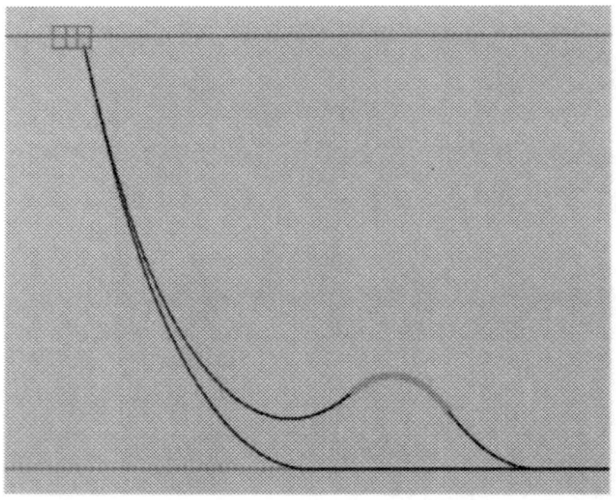

Used in water depth of 2,900m
Large diameter ~24" OD
High pressure 15 ksi

Steel Catenary Risers (SCRs)
Or
Steel Lazy Wave Riser (SLWR)

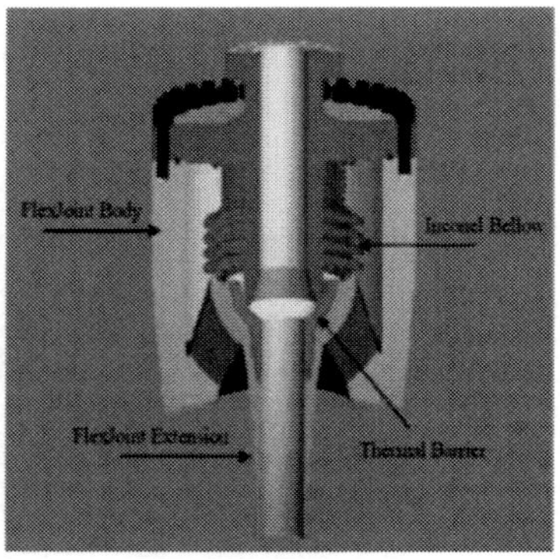

Courtesy of OilState

Flexible joint
Pressure <12 ksi
Angle < +/-25 deg
Tension <4,000 kips
Temperature -10~110

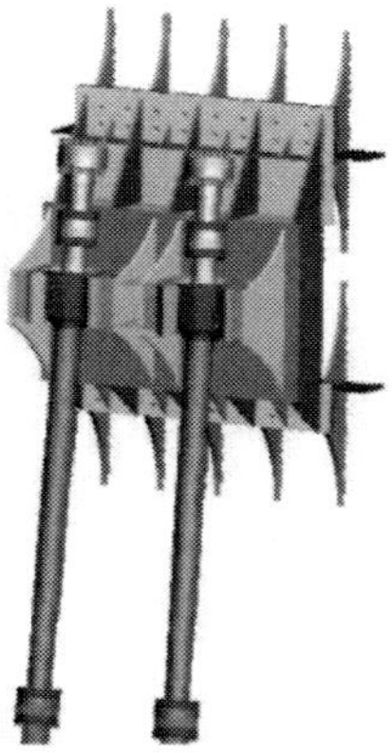

Copyright by RTI Energy Systems 2005

Stress joint
Angle <20°
Diameter 20"OD

Types of Top Termination Unit

Slide 9

Risers Types and Limitations

Courtesy of http://www.forskningsradet.no/prognett-petromaks/Nyheter/Solving_the_conundrum_of_singing_risers/1226993731411&lang=en

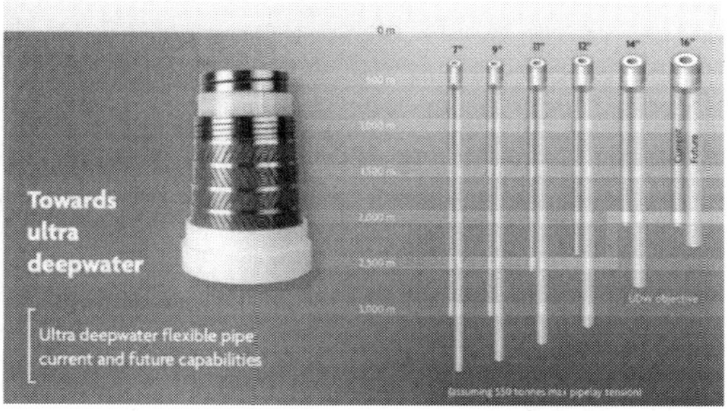

Courtesy of http://www.greencarcongress.com/2014/01/20140107-technip.html

Pressure <10 ksi
Tension limitation
Temperature limitation
Diameter 18"

Courtesy of Technip.com

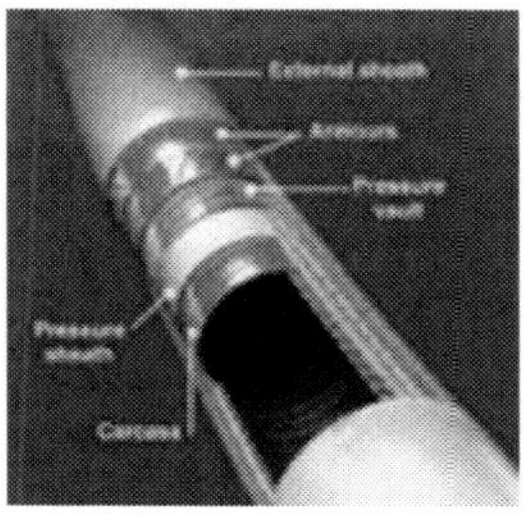

Flexible Risers

Slide 10

Risers Types and Limitations

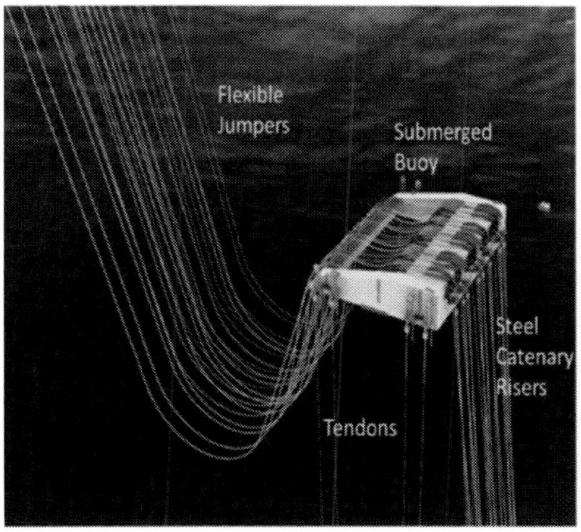

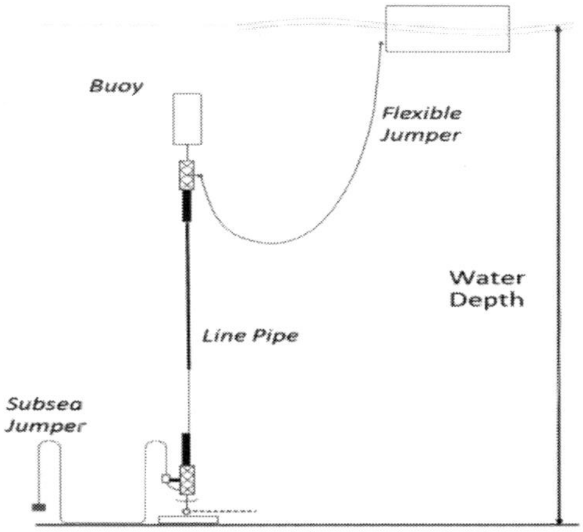

Lower steel risers & upper flexible riser
Many components, expensive, temperature limitation

Hybrid Riser

Risers Types and Limitations

	TTRs	SCRs w/Flexible Joint	SCRs w/TSJ	SLWRs w/Flexible Joint	SLWRs w/TSJ	Flexibles	Hybrid Risers
		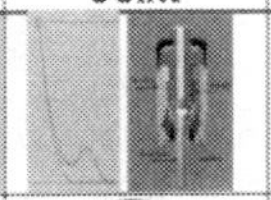	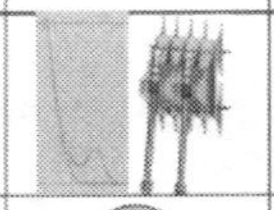	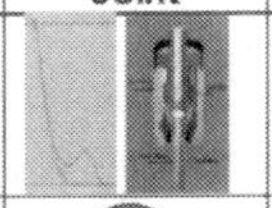	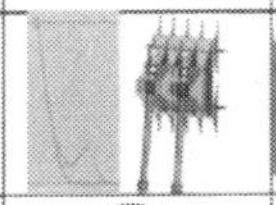		
High Pressure >15 KSI	●	●	●	●	●	●	●
High Temprature	●	●	●	●	●	●	●
Floaters with Large Dynamic Motion	●	●	●	●	●	●	●
Ultra-deepwater	●	●	●	●	●	●	●
Shallow water ~ 400 m	●	●	●	●	●	●	●
Large diameter	●	●	●	●	●	●	●
Life-cycle (OPEX) and Robustness	●	●	●	●	●	●	●
CAPEX	●	●	●	●	●	●	●

New Generation Riser Friendly Floater

The next generation floater will:

- ❑ Maintain the high oil storage capacity of conventional FPSO
- ❑ Support high deck payloads
- ❑ Superior motion response → TTR/SCR friendly
- ❑ No turret or swivel
- ❑ Large number/ large diameter risers
- ❑ Fast deployment & decommissioning
- ❑ Quayside integration
- ❑ Simple and efficient hull form
- ❑ Maintain simple topside layout (square or rectangular)
- ❑ Commercially attractive (CAPEX & OPEX)
- ❑ Low risk technology and minimum or no schedule impact

Slide 13

New Generation Riser Friendly Floater

All components are field proven

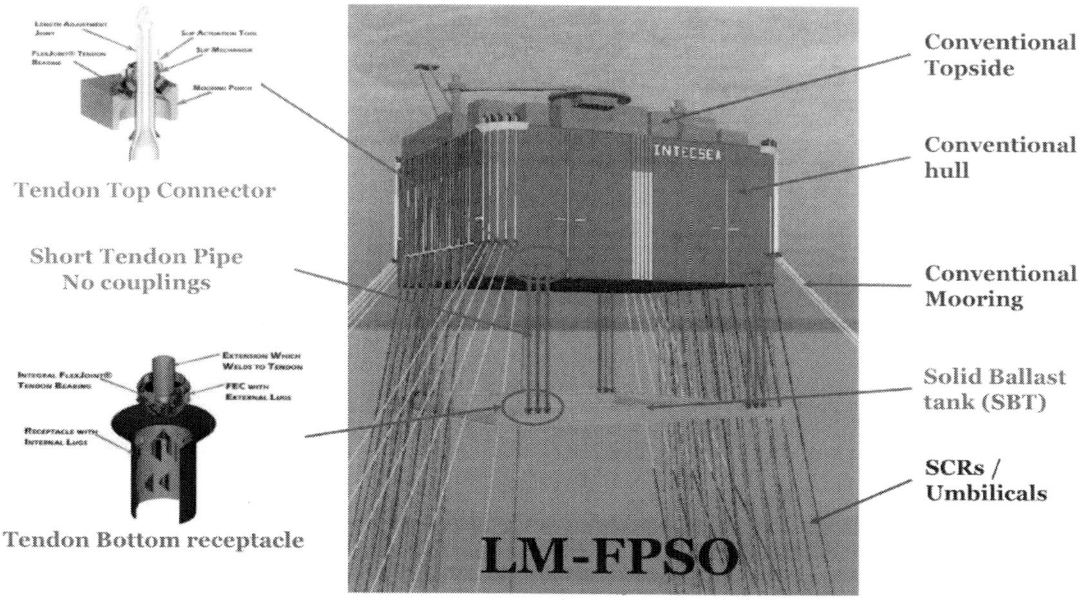

Slide 14

New Generation Riser Friendly Floater

- The mass of the SBT:
  - Maintains positive tendon tension in all design conditions
  - Ensures full coupling in heave, roll and pitch
  - Ensures full coupling in the slow motion surge/sway & yaw
  - Provides high stability (high GM) – less compartments
- SBT mass and added mass
  - Long heave, roll and pitch natural periods
  - Significantly low heave roll/pitch motions
- Relative motion between hull & SBT in surge, sway and yaw
  - Limited to first order
  - Much less than TLP hull-to-foundation relative motions

Slide 15

New Generation Riser Friendly Floater

Motion response is almost like a TLP, better than a Spar

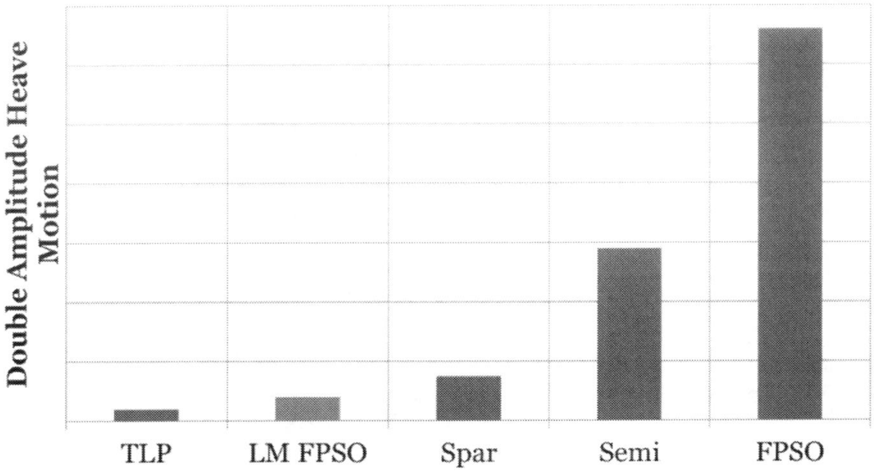

Slide 16

New Generation Riser Friendly Floater

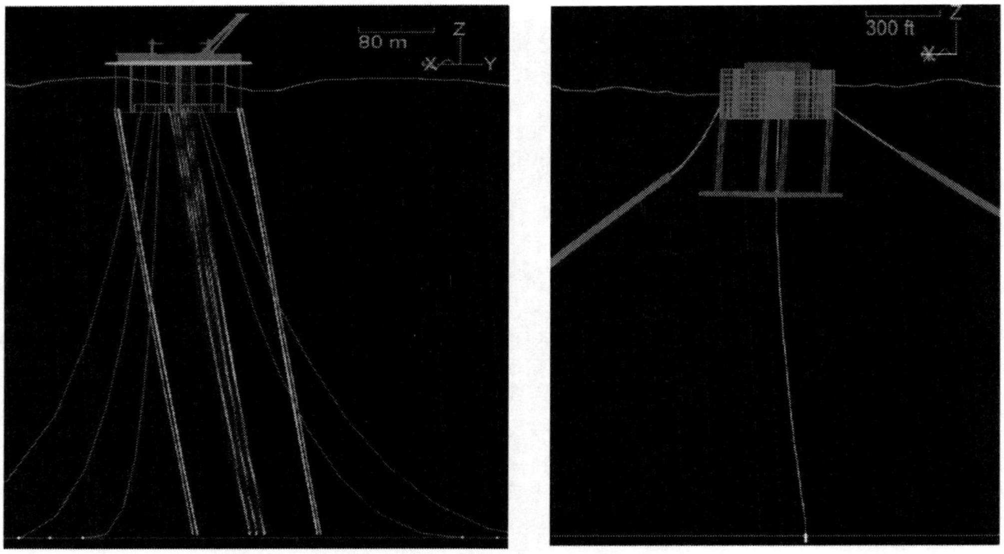

TLP vs LM-FPSO
Dry Tree & Harsh Environment
1000 yr condition

New Generation Riser Friendly Floater

The LM-FPSO has been extensively model tested in Collaboration with KRISO (Korean Research Institute for Ships and Ocean Engineering)

Case Studies: Reduced Field Development Costs

➢ A Dry-Tree Application in South China Sea:

- ➢ Water depth = 500 m
- ➢ Topside payload = 25,000 MT
- ➢ 1.0 MMBBL Storage
- ➢ 15 TTRs, 1 Multiphase production SCR
- ➢ Environment:

	1,000 Yr	100 yr
Hs (m)	15.8	13.5
Tp (sec)	16.5	15.2
Ws (m/s)	51.6	49.8
Cs @ surface (m/sec)	3.9	2.4

Slide 19

Case Studies: Reduced Field Development Costs

➢ A Dry-Tree Application in South China Sea:

	Key Figures
Hull Dimensions (m)	105 x 105 x 51.5 w/ 60 x 60 Moon-pool
Draft (m)	33.5
SBT Dimensions (m)	142 x 105 x 3.7
Tendons	16 @ 44"OD x 1.6" WT L=88 m
Natural periods (sec)	25.0 (Heave) 46.0 (Roll/pitch)
GM (m)	31

Slide 20

Case Studies: Reduced Field Development Costs

➢ A Dry-Tree Application in South China Sea:

Variable	Unit	1,000-yr	100-yr
Max SCR Porch Velocity	(m/s)	**1.7**	**0.9**
Max heave (SA)	(m)	4.0	3.1
Max combined Roll/pitch (SA)	(deg)	**3.9**	**3.4**
Max Hull-SBT rel. horizontal motion	(m)	18.8	13.5
TTR dynamic Stroke	(m)	**7.2**	**4.6**

Case Studies: Reduced Field Development Costs

- **A Dry-Tree Application in South China Sea:**

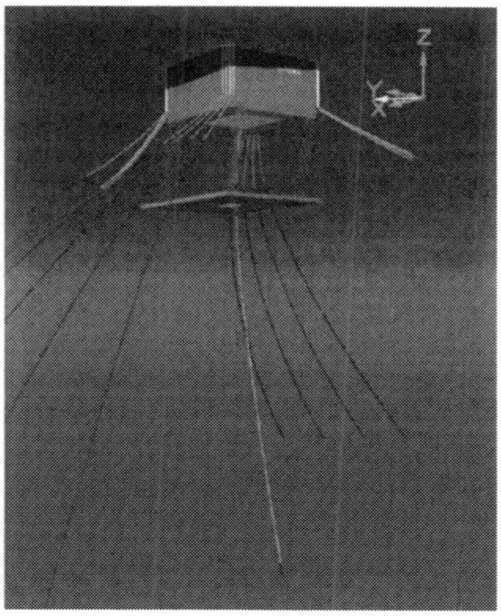

1,000-yr

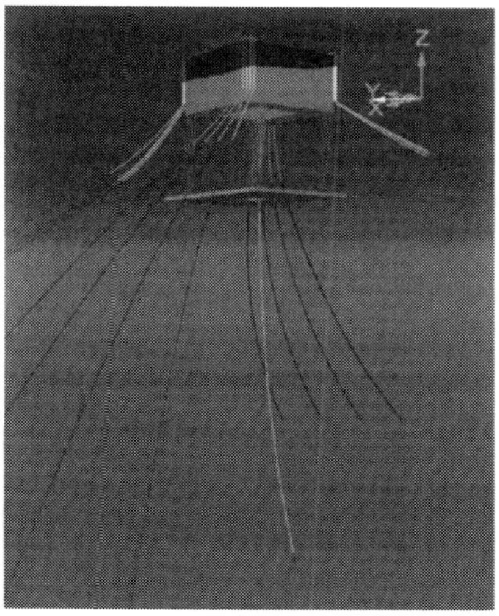

100-yr

The System Performance

Case Studies: Reduced Field Development Costs

- **A Dry-Tree Application in South China Sea:**

LM-FPSO + ONE TLP Cost Delta	
FPSO Hull and Mooring Cost Delta (TIC)	$169MM
Eliminate one TLP Hull& Mooring (TIC)	-$500MM
Topside saving	-$50MM
No turret	-$250MM
No FTL from TLP to FPSO	-$30MM
Decommissioning saving of TLP	-$100MM??
OPEX delta (1 floater vs. 2)	-$100MM??
Total Delta	**-$861 MM**

Compared to a Turret-Moored FPSO + TLP in SCS

Slide 23

Case Studies: Reduced Field Development Costs

- **A Wet-tree Application in West of Shetland:**
  - Water depth = 1,200 m
  - Topside payload = 36,000 MT
  - 1.1 MMBBL Storage
  - 22 **SCRs** + 3 umbilicals

	10,000 Yr	100 yr
Hs (m)	22.8	17.7
Tp (sec)	21.3	19.1
Ws (m/s)	45.7	37.1
Cs @ surface (m/sec)	0.5	0.5

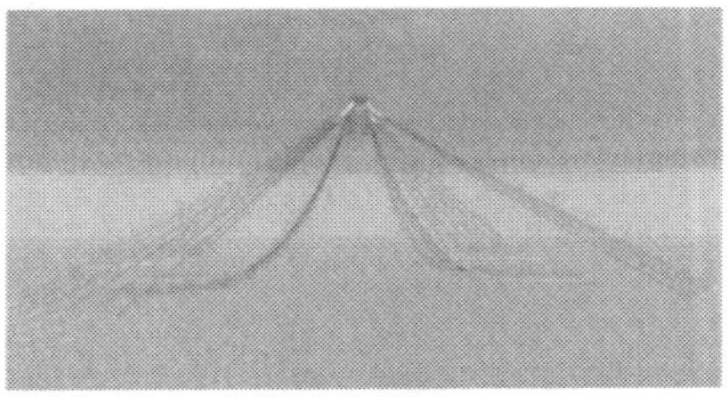

Slide 24

Case Studies: Reduced Field Development Costs

- **A Wet-tree Application in West of Shetland:**

	Key Figures
Hull Dimensions (m)	120 x 100 x 68.7 w/ 69 x 58 Moon-pool
Draft (m)	42.7
SBT Dimensions (m)	144 x 100 x 6.5
Tendons	32 @ 42"OD x 1.6" WT L=184 m
Natural periods (sec)	30.0 (Heave) 42.0/39.0 (Roll/pitch)
GM (m)	13.3

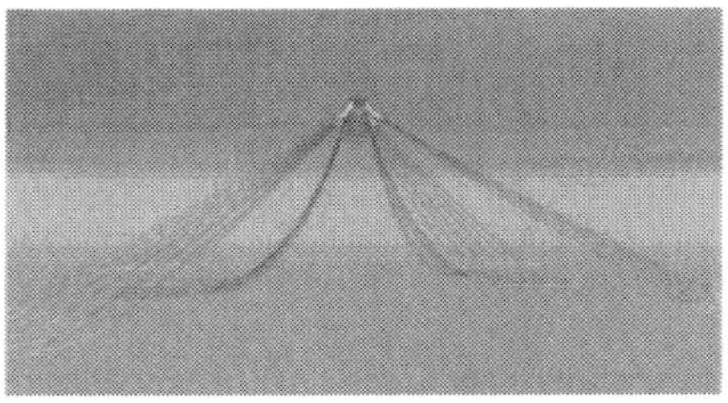

Slide 25

Case Studies: Reduced Field Development Costs

➢ A Wet-tree Application in West of Shetland:

Critical Weibull Extreme value from all headings:

Variable	Unit	10,000-yr	100-yr
Max SCR Porch Velocity	(m/s)	2.5	1.8
Max heave (SA)	(m)	7.4	4.6
Max combined Roll/pitch (SA)	(deg)	7.0	4.4
Max tendon bottom angle	(◦)	10.2	6.6

Slide 26

Case Studies: Reduced Field Development Costs

➢ A Wet-tree Application in West of Shetland:

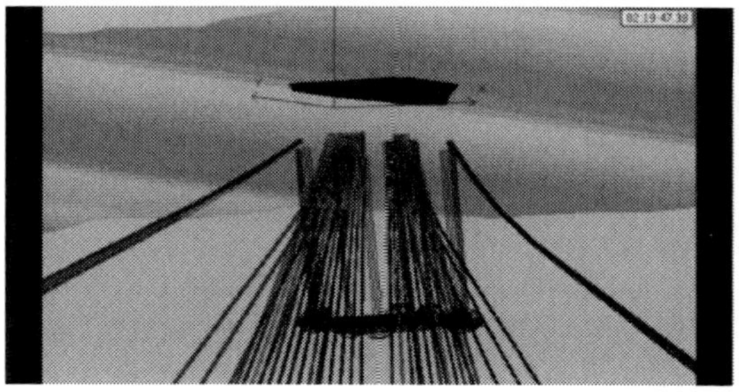

10,000-yr Critical Event

Max Relative Motion

100-yr Critical Event

Max Roll/Pitch Angle

The System Performance

Case Studies: Reduced Field Development Costs

- **A Wet-tree Application in West of Shetland:**

LM-FPSO Cost Delta	
Hull and Mooring cost delta (TIC)	$265MM
Elimination of Turret	-$750MM
22 Risers – Flexible w/Lazy wave to SCR (TIC)	-$325MM
OPEX delta (change-out of 1 flexible riser)	-$100MM??
Total Delta	**-$910 MM**

Compared to a Turret-Moored FPSO + 22 Flexible risers

Conclusions

- The LM-FPSO is a next generation floater that enables TTRs and SCRs on FPSOs
- All of the LM-FPSO components are field proven
- LM-FPSO offers improved economics by eliminating turret system and using simpler & more robust riser solution
- LM-FPSO improves platform operability with improved habitability and helicopter operation
- LM-FPSO makes project execution easier by adopting quayside integration and large GM → less VCG sensitivity during design and construction
- Case Study show that LM-FPSO saves over 900MM USD in West of Shetland wet tree application and over 850MM USD in Dry tree application in South China Sea

Slide 29

INTECSEA
WorleyParsons Group

Alaa M. Mansour, Ph.D.
Marine Engineering Manager

575 North Dairy Ashford Rd.
Houston, TX77079

1 281-206-8455 (O)
alaa.mansour@ intecsea.com

Subsea Pipeline Active Heating: Products & Flow Assurance Methodologies

Thomas Parenteau – Regional Technology Manager, Forsys Paris.

FORSYS SUBSEA

14th December 2015, Houston, TX.

1. Introduction

1. Introduction
2. Subsea Active Heating Technology Portfolio
3. Field Development Economics
4. Current Development Axis

Active Heating to Extend Solid Management (Hydrate & Wax) Options

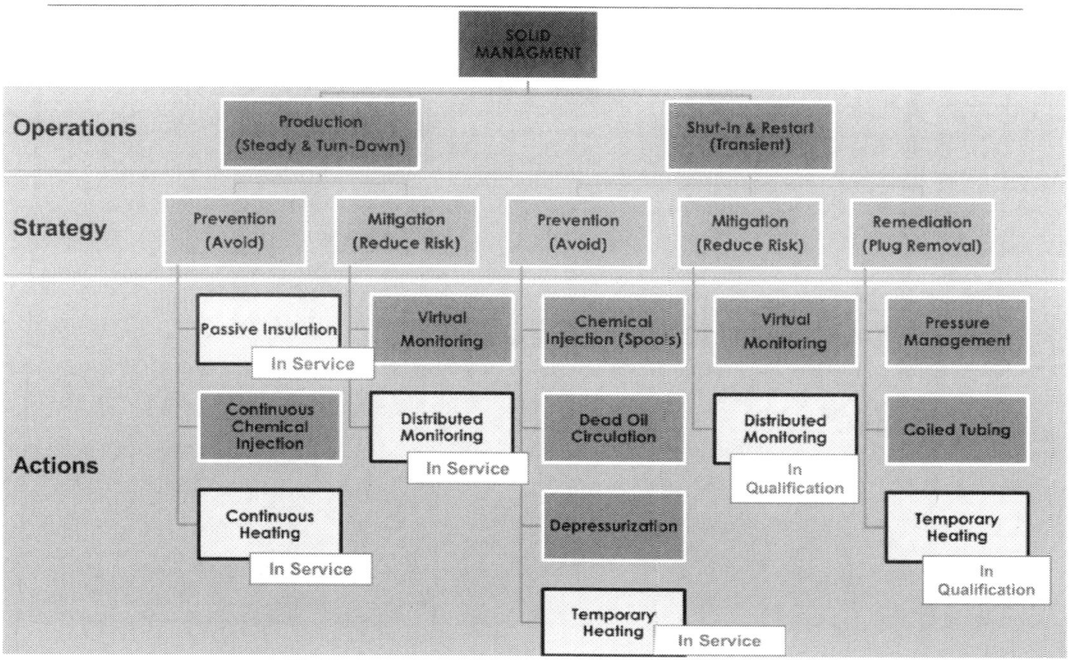

Introduction

Active Heating Demo

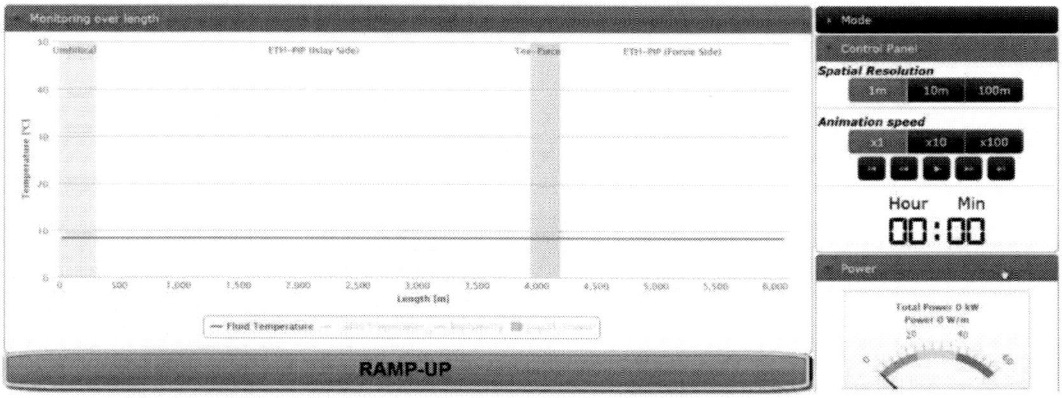

2. Subsea Active Heating Technology

1. **Introduction**
2. **Subsea Active Heating Technology Portfolio**
3. **Field Development Economics**
4. **Current Development Axis**

Subsea Active Heating Technology Portfolio

Classification by **PIPE TYPE**

	HOT WATER CIRCULATION (HW)	DIRECT ELECTRICAL HEATING (DEH)	ELECTRICAL TRACE HEATING (ETH)
PIPE IN PIPE	 HW PIP	 DEH-PIP	 ETH-PIP
WET PIPE	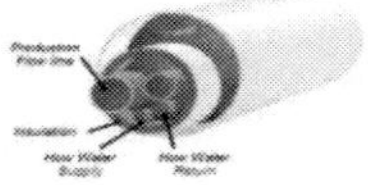 HW BUNDLE	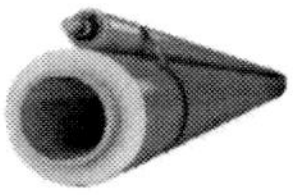 WET DEH	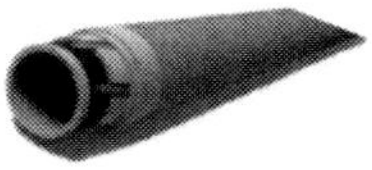 ETH-SINGLE PIPE
FLEXIBLE	 IPB	 DEH FLEXIBLE	 ETH-FLEXIBLE
INTERVENTION			 ETH-BLANKET

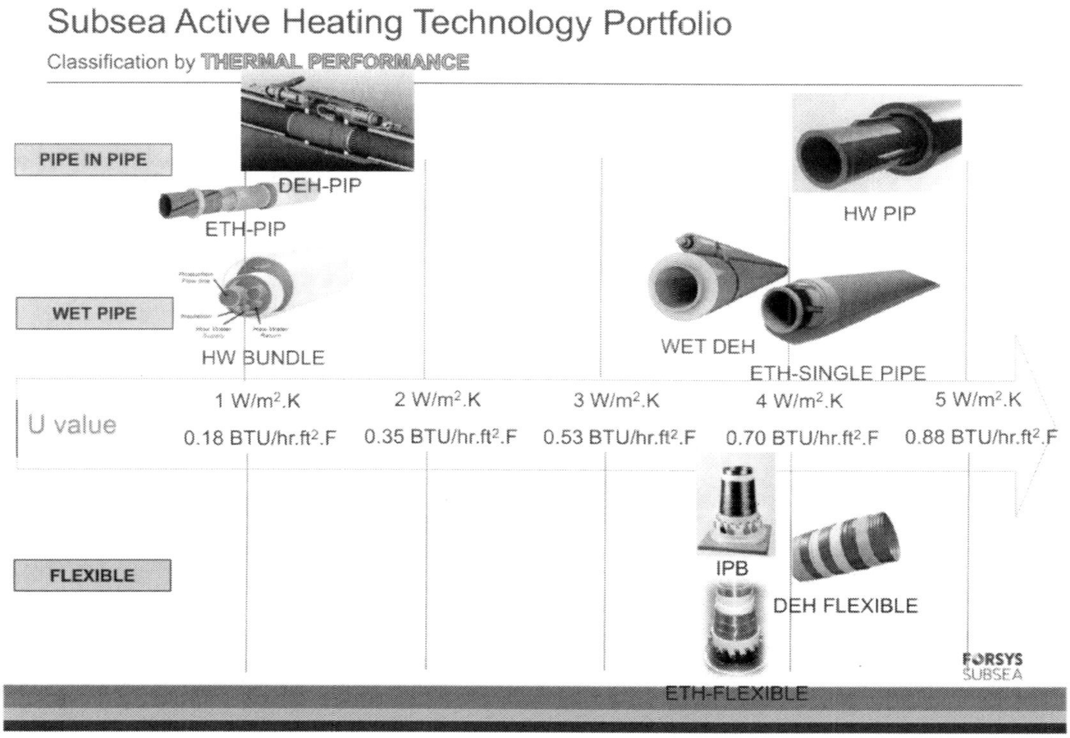
Subsea Active Heating Technology Portfolio
Classification by THERMAL PERFORMANCE
PIPE IN PIPE
DEH-PIP
ETH-PIP
HW PIP
WET PIPE
HW BUNDLE
WET DEH
ETH-SINGLE PIPE
U value
1 W/m².K
0.18 BTU/hr.ft².F
2 W/m².K
0.35 BTU/hr.ft².F
3 W/m².K
0.53 BTU/hr.ft².F
4 W/m².K
0.70 BTU/hr.ft².F
5 W/m².K
0.88 BTU/hr.ft².F
FLEXIBLE
IPB
DEH FLEXIBLE
ETH-FLEXIBLE
FORSYS SUBSEA

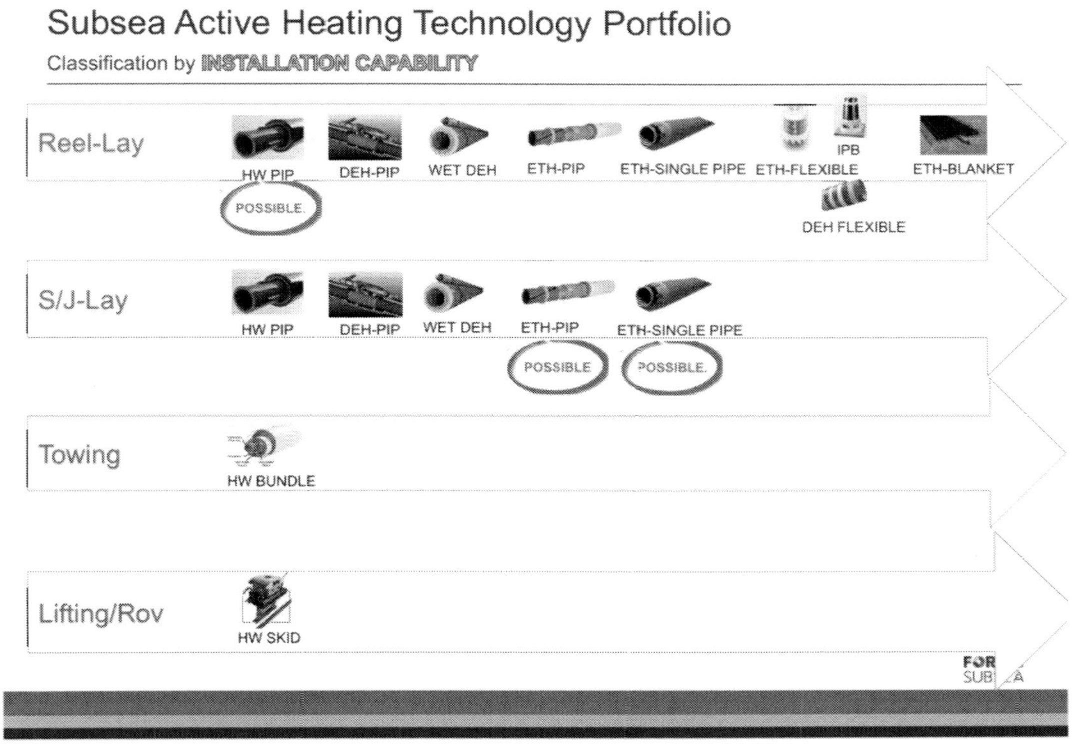
Subsea Active Heating Technology Portfolio
Classification by INSTALLATION CAPABILITY
Reel-Lay
HW PIP
POSSIBLE
DEH-PIP
WET DEH
ETH-PIP
ETH-SINGLE PIPE
ETH-FLEXIBLE
IPB
ETH-BLANKET
DEH FLEXIBLE
S/J-Lay
HW PIP
DEH-PIP
WET DEH
ETH-PIP
ETH-SINGLE PIPE
POSSIBLE
POSSIBLE
Towing
HW BUNDLE
Lifting/Rov
HW SKID

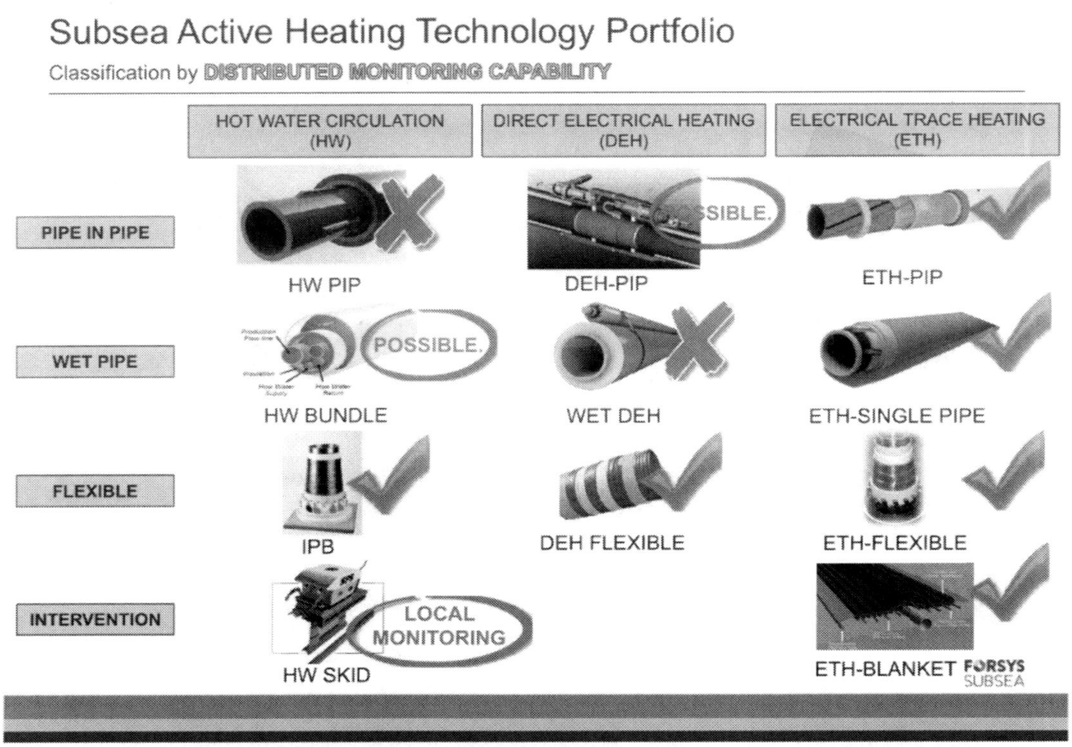
Subsea Active Heating Technology Portfolio
Classification by DISTRIBUTED MONITORING CAPABILITY
HOT WATER CIRCULATION (HW)
DIRECT ELECTRICAL HEATING (DEH)
ELECTRICAL TRACE HEATING (ETH)
PIPE IN PIPE
HW PIP
SIBLE.
DEH-PIP
ETH-PIP
WET PIPE
POSSIBLE.
HW BUNDLE
WET DEH
ETH-SINGLE PIPE
FLEXIBLE
IPB
DEH FLEXIBLE
ETH-FLEXIBLE
INTERVENTION
LOCAL MONITORING
HW SKID
ETH-BLANKET
FORSYS SUBSEA

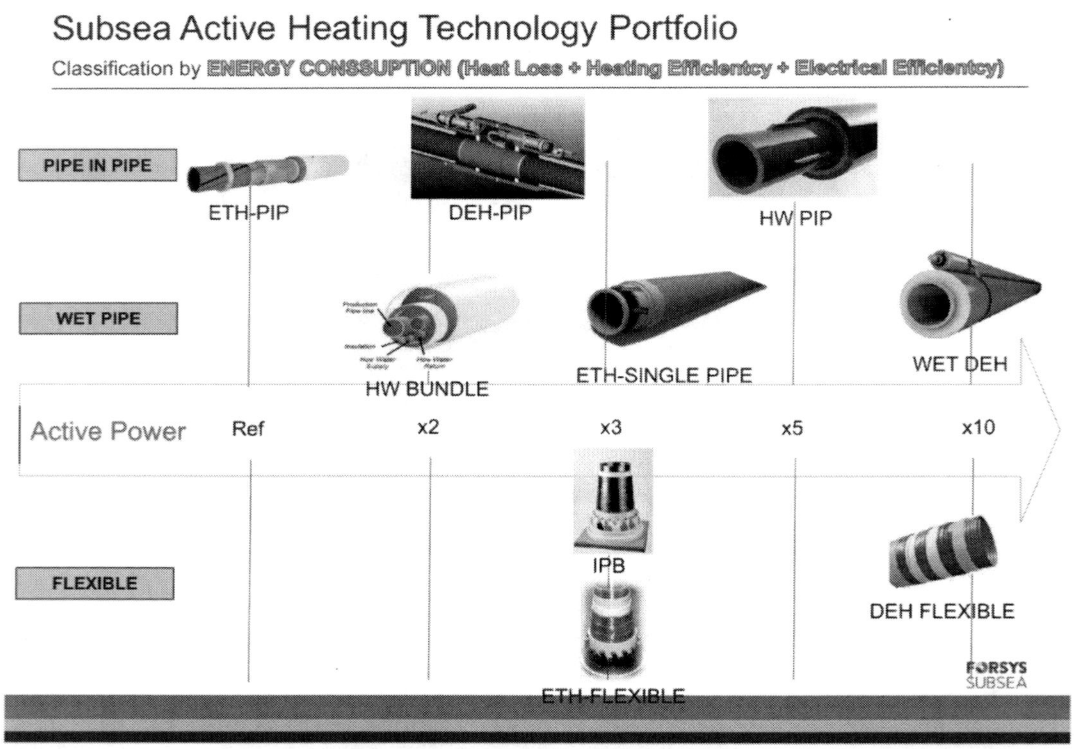
Subsea Active Heating Technology Portfolio
Classification by ENERGY CONSSUPTION (Heat Loss + Heating Efficientcy + Electrical Efficientcy)
PIPE IN PIPE
ETH-PIP
DEH-PIP
HW PIP
WET PIPE
HW BUNDLE
ETH-SINGLE PIPE
WET DEH
Active Power
Ref
x2
x3
x5
x10
FLEXIBLE
IPB
DEH FLEXIBLE
ETH-FLEXIBLE
FORSYS SUBSEA

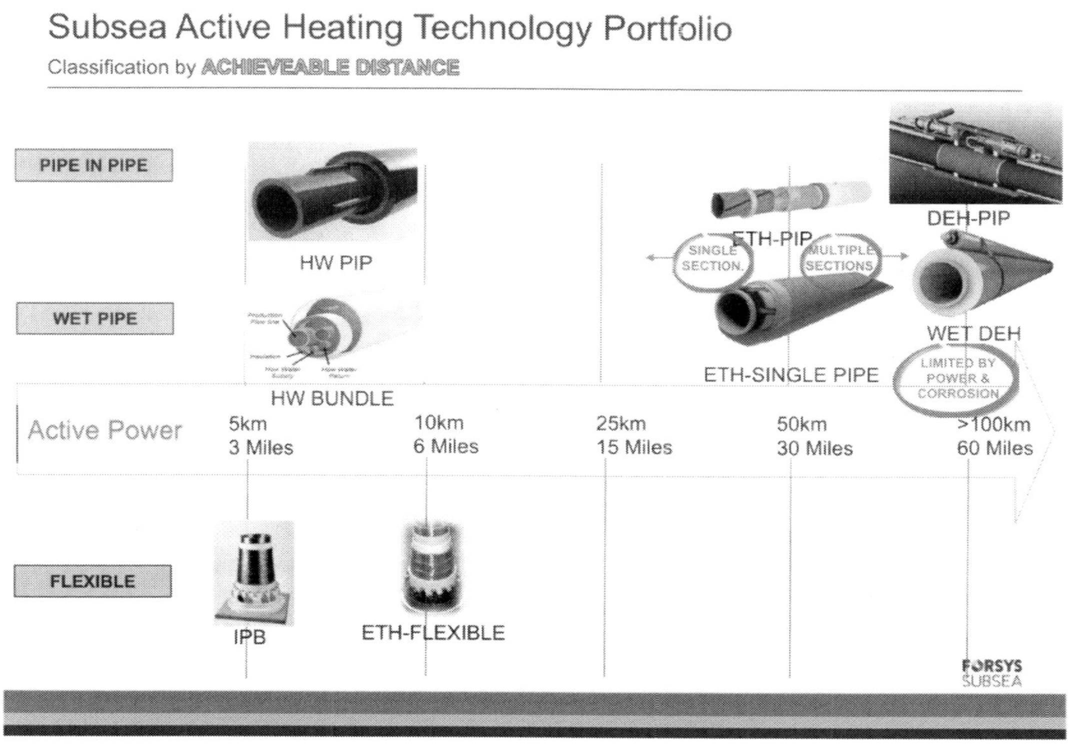
Subsea Active Heating Technology Portfolio
Classification by ACHIEVEABLE DISTANCE
PIPE IN PIPE
WET PIPE
FLEXIBLE
HW PIP
HW BUNDLE
ETH-PIP
SINGLE SECTION
MULTIPLE SECTIONS
DEH-PIP
WET DEH
ETH-SINGLE PIPE
LIMITED BY POWER & CORROSION
Active Power
5km
3 Miles
10km
6 Miles
25km
15 Miles
50km
30 Miles
>100km
60 Miles
IPB
ETH-FLEXIBLE
FORSYS SUBSEA

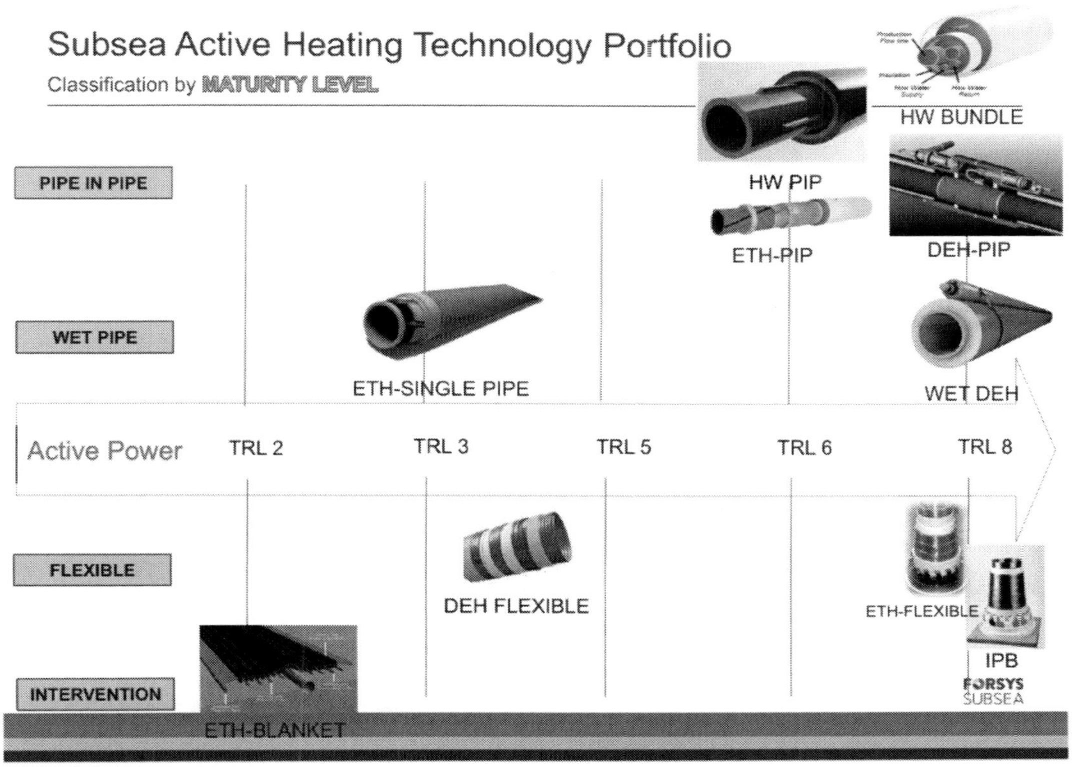
Subsea Active Heating Technology Portfolio
Classification by MATURITY LEVEL
PIPE IN PIPE
WET PIPE
FLEXIBLE
INTERVENTION
HW BUNDLE
HW PIP
ETH-PIP
DEH-PIP
ETH-SINGLE PIPE
WET DEH
Active Power
TRL 2
TRL 3
TRL 5
TRL 6
TRL 8
DEH FLEXIBLE
ETH-FLEXIBLE
IPB
ETH-BLANKET
FORSYS SUBSEA

3. Field Development Economics

1. Introduction
2. Subsea Active Heating Technology Portfolio
3. Field Development Economics
4. Current Development Axis

FORSYS SUBSEA

GoM Deepwater, HP Case Study

Concept Definition & Operating Window

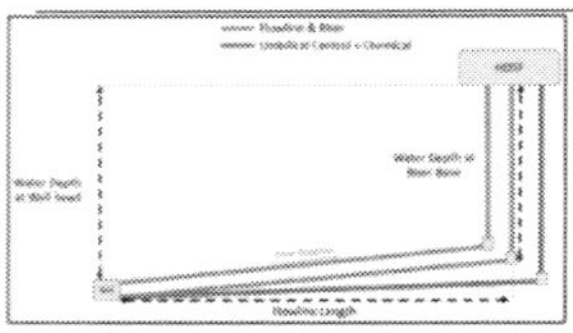

PASSIVE WET SYSTEM

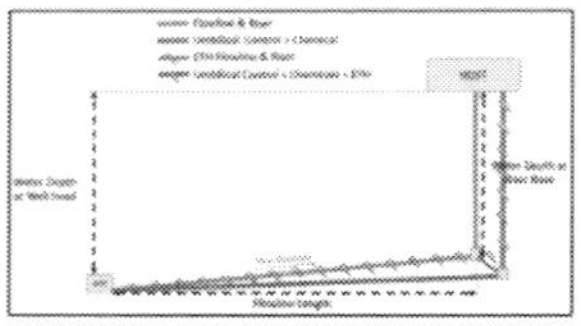

ACTIVE ETH-PIP SYSTEM (WET RISER)

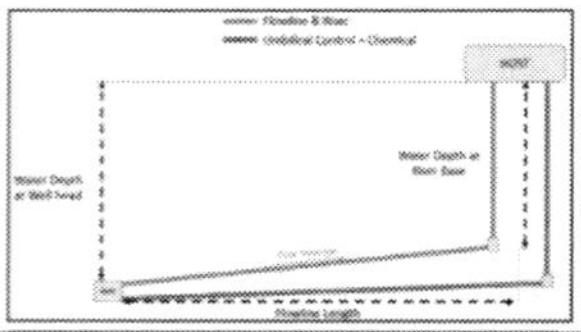

CHEMICAL WET SYSTEM

Parameter	Value
WATER DEPTH	6000 ft
LENGTH	15 Miles
DESIGN PRESSURE	12 000 Psi
TEMPERATURE	200 °F
DENSITY	20
VISCOSITY	5cP
GOR	1000
WHFP	7000 to 9000 Psi
WHFT	130 to 160 °F
WAT	100 °F
HAT	70 °F

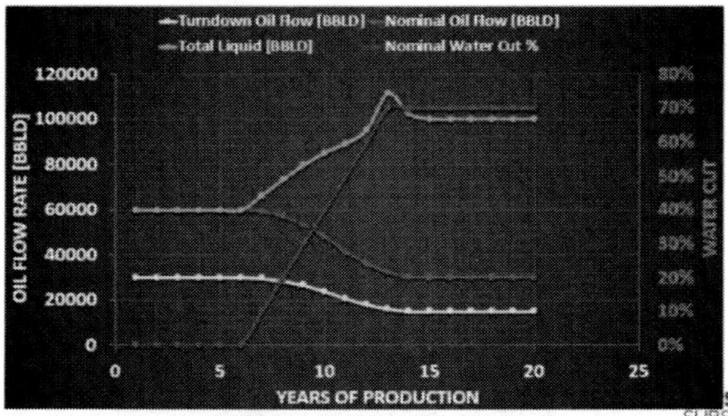

PRODUCTION PROFILE

FORSYS SUBSEA

GoM Deepwater, HP Case Study

Active Heating Technology Screening

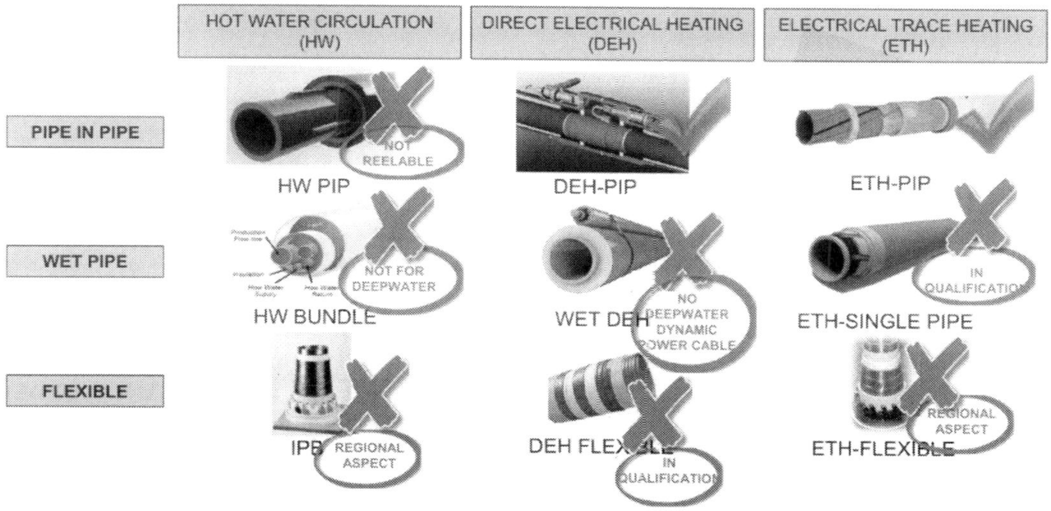

FORSYS SUBSEA

GoM Deepwater, HP Case Study

System Cost Ranking covering Procurement, Installation, Fabrication, Engineering

SURF CAPEX	REF	-20%	-15%	-45%

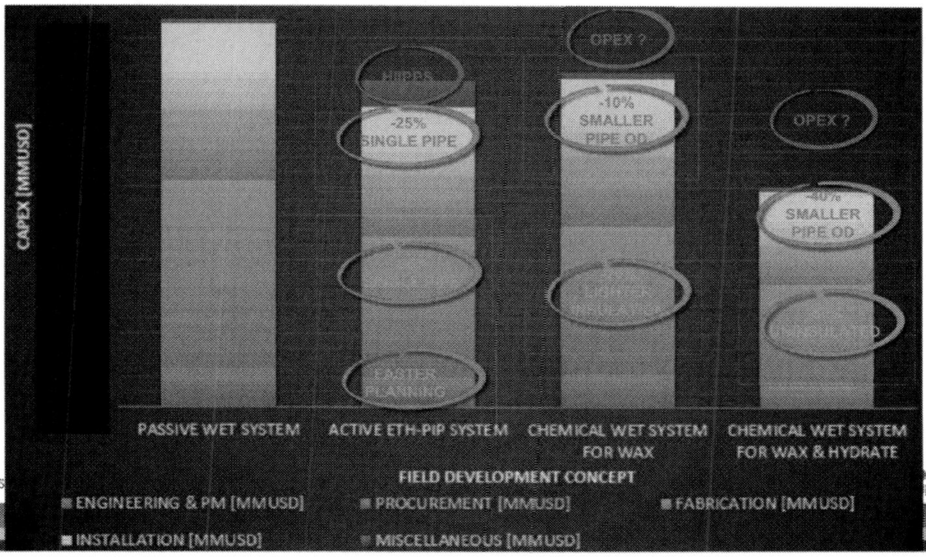

16

FORSYS SUBSEA

GoM Deepwater, HP Case Study

System Cost Ranking covering Procurement, Installation, Fabrication, Enegineering

CAPEX	REF	-20%	-15%	-45%

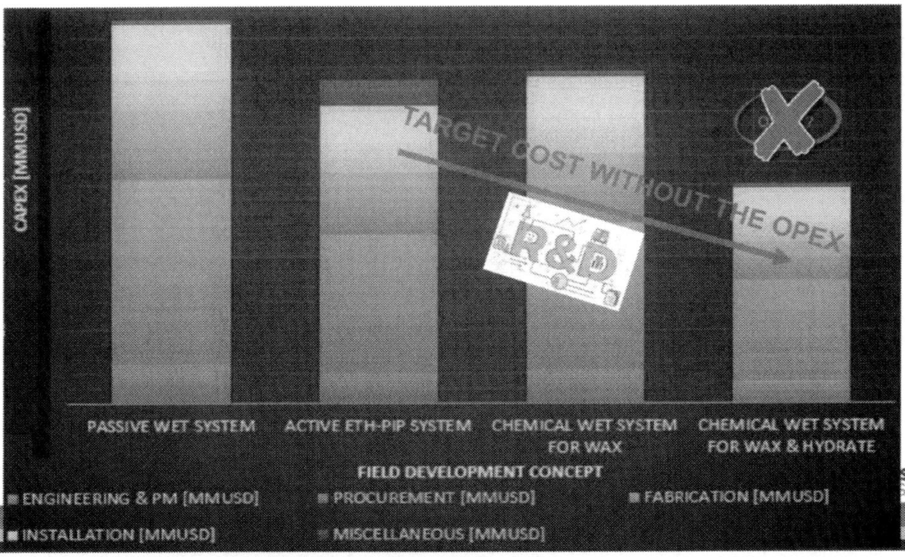

17 FORSYS SUBSEA

4. Current Development Axis

1. Introduction
2. Subsea Active Heating Technologies
3. Operability & Field Development Strategies.
4. Current Development Axis.

18 FORSYS SUBSEA

Active Heating Development for Intensive CAPEX Savings In Deepwater Environment.

- **ENABLE LONGER & DEEPER TIE-BACK THROUGH LONG RANGE ACTIVE HEATING.**
- **UNLOCK LIGHTER SYSTEMS WITH CONTINUOUS HEATING**
- **UNLOCK CHEAP SUBSEA INTERVENTION SYSTEMS TO SOLVE « RARE » WAX & HYDRATE PLUGGING ISSUES**

19

FORSYS SUBSEA

Active Heating Development for Intensive CAPEX Savings In Deepwater Environment.

- **ENABLE LONGER & DEEPER TIE-BACK THROUGH LONG RANGE ACTIVE HEATING.**

DIRECT ELECTRICAL HEATING (DEH)

ELECTRICAL TRACE HEATING (ETH)

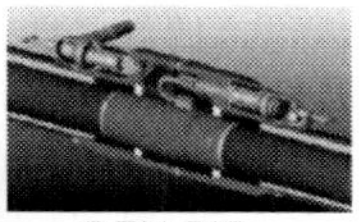

DEH-PIP

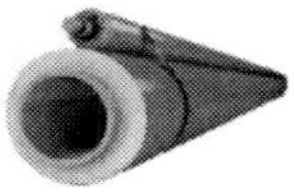

WET DEH

ETH-PIP

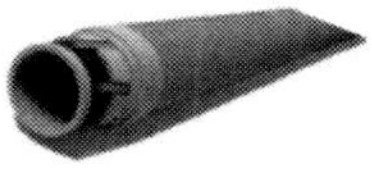

ETH-SINGLE PIPE

Qualification of Deepwater Dynamic Power Cable

Qualification of Higher Voltage ETH Cable

20

FORSYS SUBSEA

Active Heating Development for Intensive CAPEX Savings In Deepwater Environment.

- **UNLOCK LIGHTER SYSTEMS WITH CONTINUOUS HEATING**

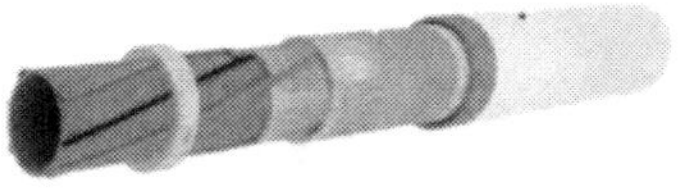

ETH-PIP

Lighter

Larger OD

ETH-SINGLE PIPE

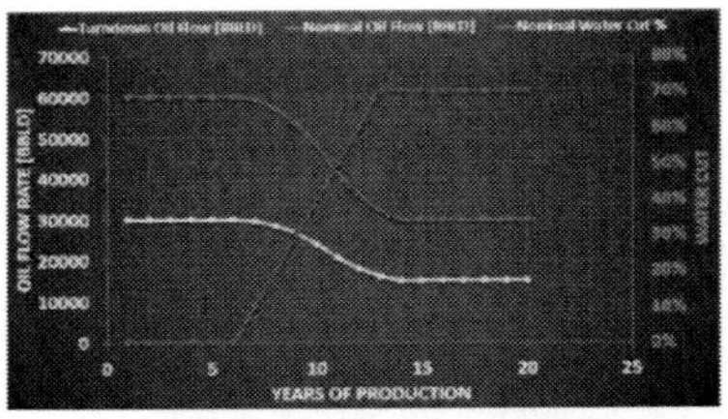

PASSIVE SYSTEM DESIGN BASIS

Continuous Heating

Insulate Only for Heating Efficiencty.

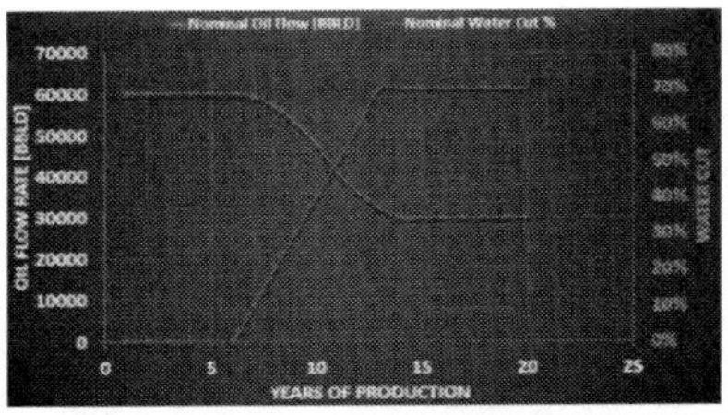

CONTINUOUS HEATING SYSTEM DESIGN BASIS

Active Heating Development for Intensive CAPEX Savings In Deepwater Environment.

- **UNLOCK CHEAP SUBSEA INTERVENTION SYSTEMS TO SOLVE « RARE » WAX & HYDRATE PLUGGING ISSUES**

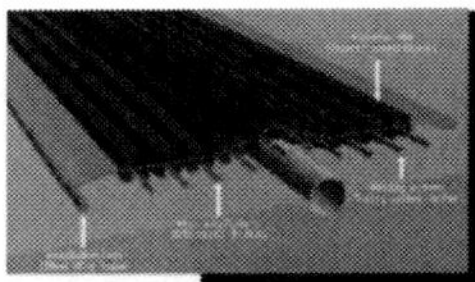

ETH-BLAN[redacted]

Active Heating Development for Intensive CAPEX Savings In Deepwater Environment.

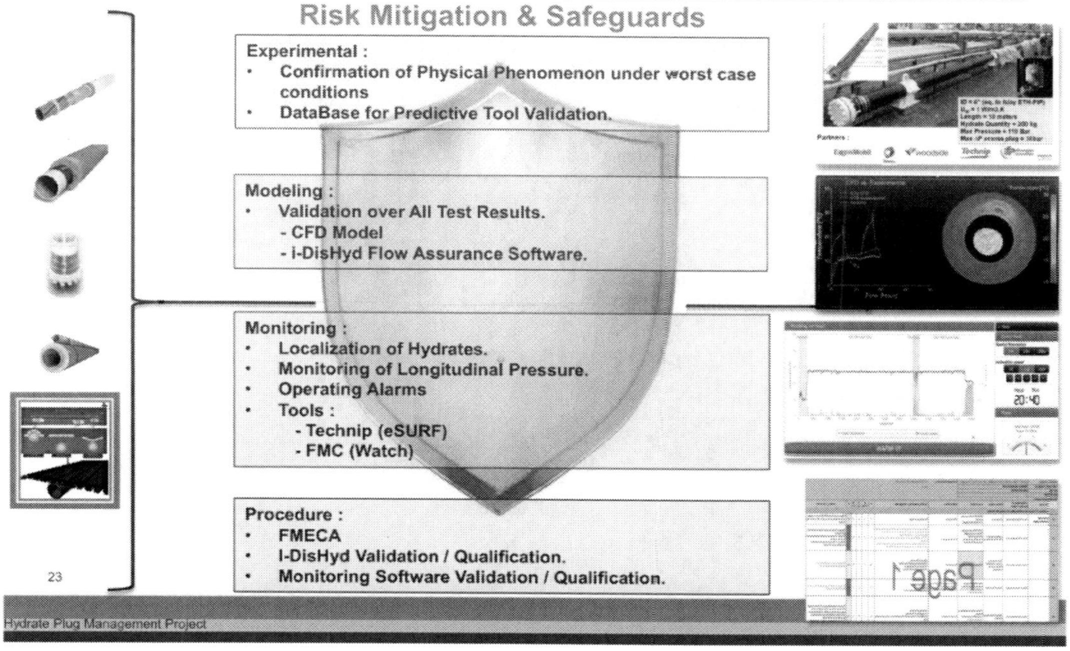

i-DisHyd Application – Pressure Profil

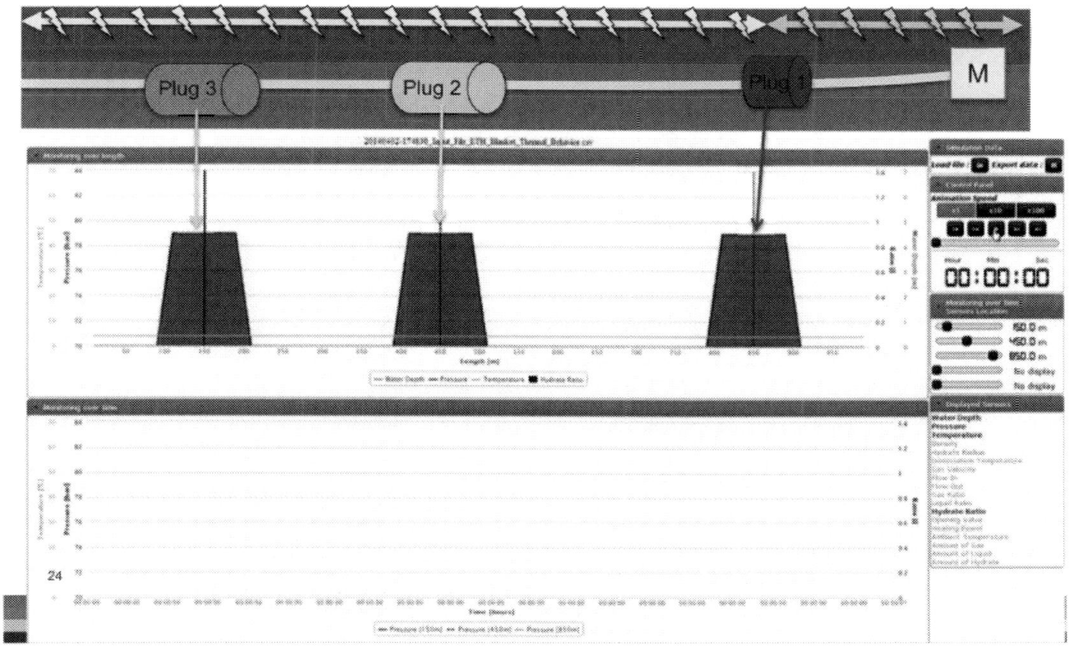

Thank You !

25

FORSYS SUBSEA

AUTHOR INDEX

AIChE
120 Wall Street, FL 23
New York, NY 10005-4020

ISBN 978-1-5108-6373-6

CURRAN ASSOCIATES INC.
proceedings.com